Oxford Chemistry Serie

General Editors
P. W. ATKINS   J. S. E. HOLKER   A. K. HOLLIDAY

Oxford Chemistry Series

K. A. McLauchlan: *Magnetic resonance* (1972)

R. J. Puddephatt: *The periodic table of the elements* (1972)

J. Robbins: *Ions in solution* (2): *an introduction to electrochemistry* (1972)

RICHARD A. JACKSON

Mechanism:

an introduction to the study of organic reactions

Clarendon Press · Oxford · 1972

Oxford University Press, Ely House, London W.1

GLASGOW   NEW YORK   TORONTO   MELBOURNE   WELLINGTON
CAPE TOWN   IBADAN   NAIROBI   DAR ES SALAAM   LUSAKA   ADDIS ABABA
DELHI   BOMBAY   CALCUTTA   MADRAS   KARACHI   LAHORE   DACCA
KUALA LUMPUR   SINGAPORE   HONG KONG   TOKYO

PRINTED IN GREAT BRITAIN BY
J. W. ARROWSMITH LTD., BRISTOL, ENGLAND

To my parents

Editor's foreword

IT has become all too common in organic chemistry for a reaction to be rationalized by drawing a series of formulae with "curly arrows" in appropriate directions. This is frequently done whether or not the mechanism of the reaction has been investigated. Unfortunately, too many students and researchers regard the "curly arrow" picture as a satisfactory answer and they seem to be unwilling to pose the further questions which this type of rationalization should suggest. This book clearly demonstrates the need to establish reaction mechanisms and it provides an understanding of the range of methods available. The reactions chosen to illustrate the use of the individual methods cover an extensive range of organic reaction types.

The text should be readily understood by second-year undergraduates but further insight will be gained by reading other books in the series. A full understanding of the use of the methods of proton magnetic resonance and electron spin resonance as mechanistic tools requires a basic knowledge of the underlying principles. These will be found in the volume on magnetic resonance. A more complete picture of the use of stereochemical methods will be obtained from the forthcoming book on stereochemistry and mechanism, and a later book on fast reactions will illustrate the specialized techniques that are necessary for exploration of this important field.

J.S.E.H.

Preface

CHEMISTRY is frequently represented as a subject which is essentially factual, and without much intellectual content. While it is true that a large body of factual knowledge exists in chemistry, the intellectual interest in the subject lies in the building up of generalizations and principles which tie together and explain the body of facts, and may in turn suggest new experiments and ideas for testing.

This process may be seen in the study of organic reaction mechanisms, illustrated in this book by a number of specially chosen reactions. No attempt has been made to cover the mechanism of every important reaction, and many important areas, such as photochemistry and heterogeneous reactions, are not considered. The tools of the organic chemist in this field are presented with comments on their use and limitations. A number of illustrative problems have been included, and it is hoped that readers will consider how the methods used in this book can be used to derive information about other reactions which they will meet. They may also be encouraged to wonder (depending on their major subject interest) which of these methods may be of use in inorganic, polymer, heterogeneous, or biochemical systems, and to compare the methods described in this book with the specialized techniques which have been developed in other areas.

Chemistry as an evolving subject has involved personalities as well as experimentation. Progress has sometimes been helped, sometimes hindered by ideas. Questions of cost and availability of apparatus (and man-power) influence the experiments which are carried out. This interplay, while not forming the main subject of the book, cannot be entirely omitted since this book is about methods and ideas as well as facts.

Brighton
November 1971   R.A.J.

Acknowledgements

I WISH to record my gratitude to the teachers who awakened and fostered my interest in Chemistry: Professor R. P. Bell F.R.S., Professor W. A. Waters F.R.S., and the late N. M. Irvin Esq.

This book owes a great deal to my colleagues at the University of Sussex, in particular Professor C. Eaborn F.R.S., Dr. M. H. Ford-Smith, Dr. F. McCapra, and Dr. J. G. Stamper, whose detailed comments on my first draft have contributed greatly to the final version. Finally, my thanks are due to a number of secretaries and to my wife for a great deal of work in connection with the production of the typescript and diagrams.

Contents

What is a reaction mechanism? Why study reaction mechanisms? Proof and reasonableness in reaction mechanisms. The establishment of organic reaction mechanisms. Problems.

Establishment of the structure of products and their yields. Which bonds have been formed and broken during the reaction? Have groups or atoms moved from molecule to molecule during the reaction? Evidence from by-products and mixtures of products. Summary. Problems.

Experimental techniques. The connection between mechanism and kinetics. Primary kinetic isotope effect. Summary. Problems.

Carbonium ions. Carbanions. Radicals. Carbenes. Benzynes. Tetrahedral intermediates. Acid-base catalysis. Stable molecule intermediates. Conclusion. Problems.

Substitution at tetrahedral centres and optical isomerism. Inversion reactions at carbon. Substitution with retention. Reactions which involve carbonium ions, radicals, or carbanions as intermediates. Additions to multiple bonds and eliminations. Concerted molecular reactions. Interpretation of new data. Problems.

Steric acceleration and deceleration. Substituent effects. Electronic effects of substituents: the Hammett equation. Solvent effects. Hammond's Postulate. Problems.

Problems.

Symbols

—✗→	Reaction which does not go.
→ →	Emphasizes that the reaction in question involves more than one step.
⇌	Reversible reaction.
⇌	Reversible reaction; equilibrium favours products.
—σ→	Reaction with inversion of stereochemistry.
⌢	Pictorial representation of the movement of a pair of electrons during a reaction from the place denoted by the tail of the arrow to the place denoted by the head.
↔	Used between two or more structures to imply that these are 'resonance' or 'canonical' forms which contribute to the stability of the molecule but do not have an independent existence.
—	Covalent bond.
⋯	Bond formed or broken during a reaction.
⋯⋯	Bond to an atom or group behind the plane of the paper.
*	Nearly always used in this book to denote an asymmetric atom, for example a carbon atom with four different substituents. Also used to denote ^{14}C isotopic substitution in aromatic rings only. All other isotopic substitutions are denoted as D, T, ^{14}C, ^{18}O, etc.

1. Reaction mechanisms—
what and why

What is a reaction mechanism?

MOLECULES consist of atoms arranged in definite positions relative to each other. The arrangement is normally one of stable equilibrium: small displacements of any of the atoms produce restoring forces which tend to pull them back to their equilibrium positions. Spectroscopic and diffraction techniques often give the distances between atoms in molecules (bond-lengths) to ± 1 pm and bond angles to within $\pm\frac{1}{2}°$.

If relatively large displacements are made of one or more atoms in one or more molecules, the molecules may not recover their initial structures; instead the atoms may take up new stable configurations in new molecules, and a chemical reaction has taken place. To produce large displacements, and hence chemical reaction, energy is required in excess of the average possessed by all molecules at a particular temperature. At any particular temperature, collisions between molecules will produce a range of molecular energies, and some molecules may acquire enough energy to react. The higher the temperature the higher the average energy, and the greater the proportion of molecules with greater than the critical energy required for reaction. Energy can be supplied in other ways, the most important being visible or ultraviolet light, which when absorbed by a molecule gives rise to electronic excitation.

To understand completely the *mechanism* of an organic reaction, we require to know as a function of time the precise positions of the atoms in the reactant molecules as they are converted into products via any possible intermediates. This is a *goal* which can never be completely achieved. The techniques of spectroscopy and diffraction, which allow the determination of chemical structure, are not well suited to follow changes of structure which take place during chemical reactions within about 10^{-13} to 10^{-14} second, that is times roughly comparable with those required for a molecular vibration or a molecular collision. Thus our knowledge of reaction mechanisms has to be based on indirect evidence. Particular pieces of evidence can disprove a particular mechanism, but they can rarely prove one. Mechanisms become established if a wide variety of evidence points to a particular mechanism, but new evidence may cause an established mechanism to be modified or abandoned.

Why study reaction mechanisms?

The utility of a chemical reaction does not depend on our understanding it. Many chemical processes such as combustion, the extraction of metals from their ores, and soap manufacture, were being carried out successfully long before the concept of reaction mechanism or even of chemistry had emerged. There are at least four principal reasons for studying reaction mechanisms.

(1) *Optimization*

The yields of organic reactions are often optimized (if at all) on a trial-and-error basis, but in some cases a knowledge of the mechanism and/or the rates of the individual reactions involved may help to show the best conditions for optimizing yields of a particular product.

For example, 1-alkenes can add on hydrogen bromide by two distinct processes, to yield the 1-bromoalkane or the 2-bromoalkane. A great deal is known about these processes, one of which involves free-radical intermediates, the other ionic species. On the basis of this information, it is possible to specify conditions which will produce almost quantitative yields of either of the two possible products.

$$R \cdot CH = CH_2 + HBr \quad \xrightarrow[\text{absence of peroxides}]{\text{very pure olefin}} \quad R \cdot CHBr \cdot CH_3 \qquad (1)$$

$$\xrightarrow[\text{other radical source}]{\text{trace of peroxide or}} \quad R \cdot CH_2 \cdot CH_2 Br \qquad (2)$$

These reactions are considered in more detail in Chapter 4.

As an industrial example of a mechanistic approach to problems of optimization, the experience of I.C.I. in the field of hydrocarbon chlorination may be quoted. An investment of £10 000 on a study of the kinetics of the reactions involved led to a decrease in by-product formation and a 40 per cent increase in output. The increased capacity allowed capital expenditure of £400 000 on new plant to be avoided.

(2) *Correlation*

Investigation of organic reaction mechanisms reveals similarities between different reactions. For example, primary and secondary alkyl halides react with hydroxide ions to produce the alcohol and a halide ion. Various lines of evidence described in this book suggest that under certain conditions this reaction is bimolecular (that is, the reaction involves one molecule each of alkyl halide and OH^-), and involves simultaneous formation of the $C-OH$ and breaking of the $C-Hal$ bonds (reaction 3).

$$HO^- + {}^{\delta+}C \overset{H}{\underset{R}{\overset{|}{-}}} \overset{\delta-}{Hal} \rightarrow HO \cdots \overset{H}{\underset{R}{\overset{|}{C}}} \cdots Hal \rightarrow HO - \overset{H}{\underset{R}{\overset{|}{C}}} + Hal^- \qquad (3)$$

The structural feature of the hydroxide ion which enables it to react in this way is an unshared pair of valence electrons which can be used to form a covalent bond. (Reagents which contain this feature are called nucleophiles.) Many nucleophiles are negatively charged, but a number of neutral molecules such as H_2O and NH_3 are also nucleophilic. It is found that other nucleophiles, for example HS^-, NH_3, and CN^-, will react by substitution in a way analogous to OH^-. Furthermore, groups other than halogen may be displaced in such reactions. Hence much of our knowledge about alkyl halide chemistry may be rationalized in terms of the bimolecular nucleophilic substitution reactions (commonly referred to as S_N2 reactions) outlined above. This helps us to remember the reactions of alkyl halides and suggests other reactions which might occur and still others which may be expected not to occur.

A rather limited number of types of organic reactions are known and most of the thousands of different organic reactions can be grouped into these types, thereby bringing order into the subject.

(3) Prediction

Various predictive uses of a knowledge of organic reaction mechanisms are possible. For a particular type of reaction of known mechanism, the effects of minor changes, for example of substituents or solvents, may often be predicted, either qualitatively (i.e. as making the reaction faster or slower) or quantitatively, on the basis of information obtained from other reactions.

This type of approach has been very successful with reactions of aromatic compounds, and some examples are considered in Chapter 6. In a different area of chemistry, the addition of hydrogen bromide and other molecules to olefins is known to give saturated compounds. Elucidation of the mechanism of such additions has suggested the correct experimental conditions for carrying out more difficult addition processes, such as the addition of acetic acid to 1-octene (reaction 4).

$$CH_3 \cdot CO_2H + C_6H_{13} \cdot CH = CH_2 \xrightarrow[105°C,\ 48\ h]{(Bu^tO)_2} C_6H_{13} \cdot CH_2 \cdot CH_2 \cdot CH_2 \cdot CO_2H$$

(300 mol)   (1 mol)   (69 %, based on olefin)

$$(4)$$

An interesting example of the predictive use of a knowledge of reaction mechanisms occurs in the design of rocket engines. The thrust developed under particular conditions can be estimated from the heat liberated in the combustion process

$$\text{(e.g. } 2H_2 + O_2 \rightarrow 2H_2O; \quad CH_4 + 2O_2 \rightarrow CO_2 + 2H_2O).$$

However, these calculations over-estimate the thrust, and a better estimate is obtained by considering the combustion in terms of the individual reactions which make up the process, that is, the mechanism of the reaction. Calculations based on the rates of these individual reactions show that in the time available the combustion process is not complete, and hence the thrust will be less than the thermodynamic estimate, which is based on complete combustion.

Prediction of entirely new reactions may be possible. Our knowledge of S_N2 reactions (last section), based on a large number of examples, allows us to predict with some confidence the feasibility of new reactions of this type.

The relative importance of prediction (based on past knowledge) as against accidental discovery in advances in chemical research is difficult to assess. Both undoubtedly play an important part, but in published work there is a tendency to over-estimate the logical predictive element.

(4) *Curiosity*

There is a considerable intellectual challenge in determining, largely by indirect evidence, the detailed course of chemical reactions at the molecular level. Many chemists would regard this problem as being the heart of the subject of chemistry, and therefore particularly worthy of intensive study, quite apart from the considerable utility of such investigations.

'Proof' and 'reasonableness' in reaction mechanisms

I have indicated above that there is no such thing as absolute proof of the correctness of a reaction mechanism. A recent example of an apparently well-established reaction mechanism which has been proved to be faulty is the dimerization of triphenylmethyl radicals (considered in the next chapter), and many less spectacular cases of the abandonment or modification of reaction mechanisms could be cited. For these reactions, new evidence showed that the accepted reaction mechanism would have to be modified. What principles then govern our acceptance of a reaction mechanism? The following points guide thought on the matter.

(1) *The proposed mechanism should be as simple as possible, while still accounting for the experimental facts*

The fact that the reaction of hydroxide ions with primary alkyl halides is kinetically of first order with respect to both reagents leads us first to propose that the reaction is a bimolecular reaction, in which a hydroxide ion reacts directly with an alkyl halide molecule with direct displacement of a halide ion and formation of the alcohol. This is the simplest explanation, and in general we shall look at alternative, more complex, mechanisms only if the simple mechanism proves to be inadequate. As an example of the need for a more complex hypothesis, we may take the reaction of iodobenzene with potassium amide in liquid ammonia. This gives aniline, and the simplest hypothesis would be that a direct substitution takes place, as in (5).

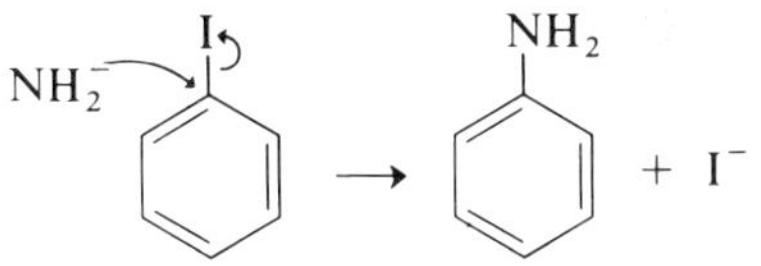

$$NH_2^- \quad \overset{I}{\underset{}{\bigcirc}} \longrightarrow \overset{NH_2}{\underset{}{\bigcirc}} + I^- \qquad (5)$$

However, experiments with the ^{14}C-labelled compound **1** show that this simple formulation is inadequate, since the incoming NH_2 group is found to enter the molecule either at the ^{14}C-labelled position or at the adjacent position, in the ratio 1:1 (within experimental error). To account for this, a two-stage mechanism involving a benzyne intermediate **2** has been proposed.

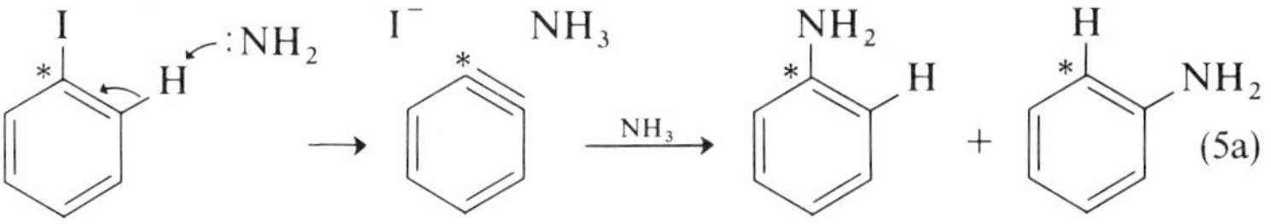

$$\underset{\textbf{1}}{\overset{I}{\bigcirc}\ :NH_2} \longrightarrow \underset{\textbf{2}}{\overset{I^-}{\bigcirc}\ NH_3} \xrightarrow{NH_3} \underset{47\%}{\overset{NH_2}{\bigcirc}\ H} + \underset{53\%}{\overset{H}{\bigcirc}\ NH_2} \qquad (5a)$$

It can be seen from the above that a reaction may be simple or *elementary*, that is it takes place in a single step without the formation of intermediates, or it may be complex, consisting of several elementary reactions.

(2) *The proposed mechanism should if possible suggest tests of its correctness*

In the above example the postulation of a reactive benzyne intermediate **2** suggests that it might be possible either to observe it spectroscopically, or to trap it by reaction with another molecule. Numerous other methods for testing particular hypotheses about mechanism are described in Chapters 2–6.

(3) *Individual steps (elementary reactions) in the proposed mechanism should be either unimolecular or bimolecular*

Unimolecular reactions occur as a result of reorganization of the bonds within a molecule, with or without rupture into fragments. The energy required for this process will have been acquired by collision with other molecules (thermal reaction) or by capture of a photon (photochemical reaction). Reactions (6)–(10) are examples of unimolecular elementary reactions which proceed thermally.

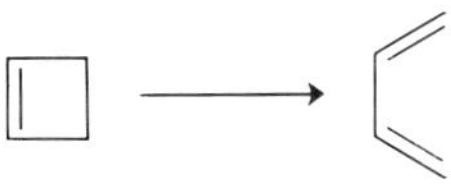

$$\square \longrightarrow \diagup\!\!\!\diagdown \qquad (6)$$

$$CH_3 \cdot CO_2 \cdot CH_2 \cdot CH_3 \rightarrow CH_3 \cdot CO_2H + C_2H_4 \tag{7}$$

$$t\text{-BuCl} \rightarrow t\text{-Bu}^+ + Cl^- \tag{8}$$

$$PhN_2^+ \rightarrow Ph^+ + N_2 \tag{9}$$

$$\mathbf{3} \qquad \mathbf{4}$$

$$CH_3-\underset{\underset{CH_3}{|}}{\overset{\overset{CH_3}{|}}{C}}-CH_2^+ \rightarrow CH_3-\underset{\underset{CH_3}{|}}{\overset{\overset{CH_3}{|}}{\overset{+}{C}}}-CH_2 \tag{10}$$

$$\mathbf{5} \qquad \mathbf{6}$$

Of these reactions, only (6) and (7) are complete in themselves. Reaction (8) would be followed by further reactions of the t-butyl cation, whereas reactions (9) and (10) would both require previous reactions to produce the cations **3** and **5**, as well as subsequent reactions to consume the cations **4** and **6**.

Bimolecular reactions take place during a collision of two molecules (which may be the same or different). At least one new covalent bond is formed between an atom from one molecule and an atom from the other molecule; additionally other bonds may be broken or formed.

$$2CF_2{=}CF_2 \quad \longrightarrow \quad \begin{array}{c} CF_2-CF_2 \\ | \qquad | \\ CF_2-CF_2 \end{array} \tag{11}$$

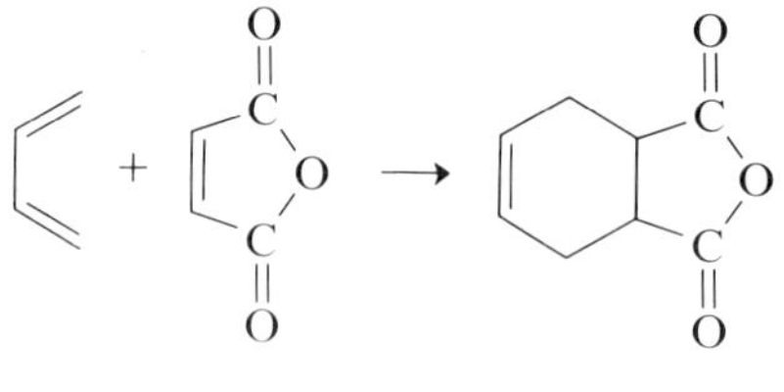

$$(12)$$

$$CH_3 \cdot + C_2H_5 \cdot \rightarrow C_3H_8 \tag{13}$$

$$HO^- + CH_3Cl \rightarrow CH_3OH + Cl^- \tag{14}$$

$$NO_2^+ + C_6H_6 \rightarrow C_6H_5 \cdot NO_2 + H^+ \tag{15}$$

Termolecular reactions would involve a collision between three molecules (with the correct orientation for reaction). Such events are very improbable and virtually rule out the possibility of termolecular reactions (a few possible examples of gas-phase inorganic termolecular reactions are known). Hence proposed mechanisms in which three or more reagents react in one step to give the observed products should be broken down into stages which involve no more than two reacting molecules.

(4) *The proposed mechanism should not violate the principle of microscopic reversibility*

This principle states that for an elementary reaction, the reverse reaction will follow the same route, but in the opposite direction. This seems intuitively reasonable, and the principle may be justified by arguments based on time-reversal. The importance of the principle of microscopic reversibility to mechanistic studies is that any information about the mechanism of an elementary reaction can be applied to its reverse reaction. An important qualitative application is that since reactions of molecularity greater than three are not encountered, reactions in which more than three molecules are produced in an elementary reaction are correspondingly implausible. For example, tetraethyl lead decomposes on heating to give lead and ethyl radicals. However, the mechanism is unlikely to involve reaction (16) as an elementary process, since its reverse would be a quinquemolecular reaction.

$$PbEt_4 \; \rightarrow \; Pb + 4\,Et\cdot \tag{16}$$

It should be noted that essentially *irreversible* reactions which produce three molecules as products are quite common in organic chemistry, even though well-established reactions involving three molecules as reactants are rare.

The conditions under which this principle can be applied to complex reactions are considered in Chapter 3.

(5) *Individual reactions should be chemically reasonable*

If intermediates are postulated, these should not offend valence rules: in particular, the first-row elements can use only 2s and 2p orbitals for bonding. The movements of atoms involved during the reaction should correspond to possible paths. Bromine adds across an olefinic double bond in a *trans* manner. This observation effectively rules out a concerted four-centre molecular mechanism (17a) in which both carbon–bromine bonds are formed simultaneously, since such a process would lead to *cis* addition. *Trans* addition (17b) requires approach of the two bromine atoms from opposite sides of the olefin (as in 7): approach in this manner is clearly inconsistent with attack on the double bond by a single undissociated molecule of bromine. Accordingly, alternative mechanisms for the addition reaction must be sought.

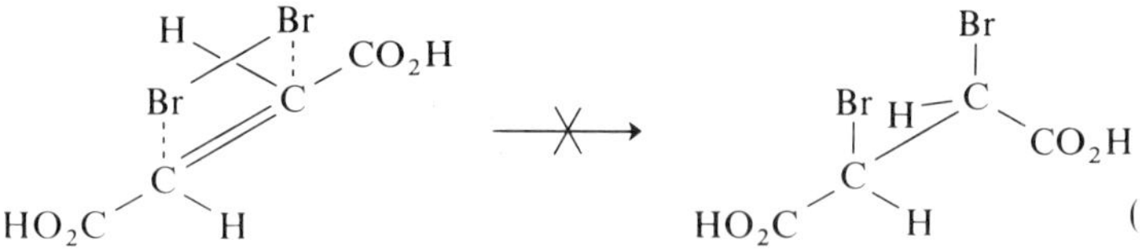

(17a)

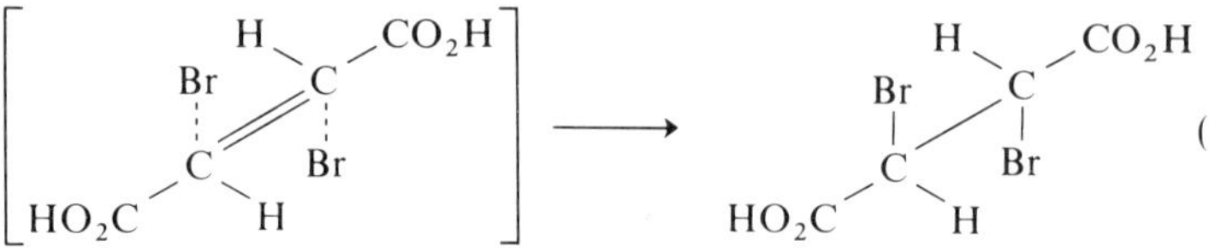

(17b)

7

observed product from fumaric acid.

In the base-catalysed condensation of benzaldehyde with acetophenone, two 'mechanisms' (18) and (19) were originally proposed for the formation of the alcohol **10** (which itself reacts further under these conditions to give the unsaturated ketone Ph·CO·CH=CH·Ph).

$$\text{Ph·CO·CH}_2\text{—H} + {}^{-}\text{OEt} \rightarrow \underset{\mathbf{8}}{\text{Ph·CO·CH}_2^-} + \text{HOEt}$$

$$\text{Ph·CO·CH}_2^- + \text{PhCHO} \searrow \underset{\mathbf{9}}{\text{Ph·CO·CH}_2\text{·CH(O}^-\text{)Ph}} \xrightarrow{\text{HOEt}} \underset{\mathbf{10}}{\text{Ph·CO·CH}_2\text{·CH(OH)Ph}} \quad (18)$$

$$\text{Ph·CHO} + {}^{-}\text{OEt} \rightleftharpoons \underset{\mathbf{11}}{\text{Ph—CH(O}^-\text{)—OEt}}$$

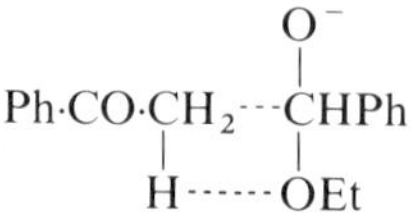

12

$$\searrow \text{EtOH} + \underset{\mathbf{9}}{\text{Ph·CO·CH}_2\text{·CH(O}^-\text{)Ph}} \xrightarrow{\text{HOEt}} \underset{\mathbf{10}}{\text{Ph·CO·CH}_2\text{·CH(OH)Ph}} \quad (19)$$

In the first mechanism (18), an ethoxide ion abstracts a proton from an acetophenone molecule to give the anion **8**, which in turn reacts with benzaldehyde by forming a bond to the carbonyl carbon to give **9**. Finally a proton is picked up from an ethanol molecule to complete the reaction and regenerate the ethoxide ion. All these processes can be visualized in terms of the movements of atoms and the formation and breaking of bonds in elementary reactions which are all analogous to other known processes. In the alternative scheme (19), the first stage which consists of addition of the ethoxide ion to a benzaldehyde molecule to give **11** is also feasible. However, it is difficult to see how the intermediate **11** could react with an acetophenone molecule to give **9** by any likely process of bond formation and rupture. Since a new carbon–carbon bond has to be formed, a transition state similar to **12** would have to be formulated, and such four-centre transition states are known to be unfavourable and require very high activation energies in other cases. Accordingly the first route (18) appears more promising as a possible mechanism, while the second mechanism (19) (which was actually preferred by the original authors on other grounds) appears improbable.

(6) *Individual reactions should be energetically reasonable*

Endothermicities, and hence minimum activation energies, can be estimated, especially for free-radical and molecular reactions. These activation energy estimates can be combined with estimates of Arrhenius pre-exponential factors (see Chapter 3) to give an upper limit for the rate of the reaction considered at a particular temperature. If the observed rate is considerably greater than this 'theoretical' value, this is good evidence that the postulated mechanism is wrong, and that another route is being followed.

Two examples illustrate this idea. Chlorination of alkanes by molecular chlorine is extremely fast, and is thought to involve a radical chain reaction with reactions (20) and (21) as the propagation steps. Values of ΔH are given for R = Me.

$$Cl_2 \xrightarrow{hv} 2Cl\cdot$$

$$R-H + Cl\cdot \rightarrow R\cdot + H-Cl \qquad \Delta H = +2\,kJ\,mol^{-1} \qquad (20)$$

$$R\cdot + Cl-Cl \rightarrow R-Cl + Cl\cdot \qquad \Delta H = -106\,kJ\,mol^{-1} \qquad (21)$$

An alternative chain-reaction mechanism would involve reactions (22) and (23).

$$Cl\cdot + R-H \rightarrow Cl-R + H\cdot \qquad \Delta H = +85\,kJ\,mol^{-1} \qquad (22)$$

$$H\cdot + Cl-Cl \rightarrow H-Cl + Cl\cdot \qquad \Delta H = -189\,kJ\,mol^{-1} \qquad (23)$$

However, reaction (22) in the second scheme is extremely endothermic and will have a very high activation energy (not less than the endothermicity of

85 kJ mol^{-1}). Given a reasonable estimate for the Arrhenius A-factor and the other variables involved, it can be calculated that reaction (22) will lead to production of chloroalkane at a rate of only about 10^{-12} mol per litre per second; put another way the half-life for the reaction will be in the region of 3000 years. Since mixtures of alkanes and chlorine explode on exposure to light (reaction complete in a fraction of a second) we may conclude that reactions (22) and (23) cannot be responsible for the chlorination process.

The reaction of fluorine with alkanes poses a different problem: fluorination takes place in the dark even at temperatures as low as $-80°C$, presumably by a radical chain mechanism similar to the one above.

The problem here is where do the radicals come from? The reaction $F_2 \rightleftharpoons 2F\cdot$ is 158 kJ mol^{-1} endothermic and can be calculated to produce very few radicals at $-80°C$. Hence the initial radicals must be produced by a different reaction; and the most likely candidate is the bimolecular reaction (24) which is only 24 kJ mol^{-1} endothermic.

$$CH_3-H + F-F \rightarrow CH_3\cdot + H-F + F\cdot \qquad \Delta H = +24\,kJ\,mol^{-1} \quad (24)$$

(7) *The mechanism should be in line (usually) with what is known about analogous reactions*

If for example, the mechanism for the hydrolysis of n-propyl acetate had been studied carefully under a particular set of conditions, it would be entirely reasonable to suggest that n-butyl acetate would react by a similar mechanism (and at approximately the same rate) under the same conditions. Indeed, a considerable body of evidence (together with some rationalization) would be needed to convince most chemists that these reactions were different. However, the t-butyl compound might very well react differently from the n-butyl compound, by analogy with a host of other known reactions of t-butyl, n-butyl, and n-propyl compounds. On a more general level, reaction mechanisms which would violate well-established principles of organic chemistry such as Bredt's bridgehead rule (double bonds are not formed at the bridgehead positions of bridged small ring systems) would be regarded with suspicion. However a slavish reliance on precedents and analogy is dangerous, since entirely new types of compound and reactions are occasionally discovered. Gomberg's discovery in 1900 that triphenylmethyl chloride reacts with metals to give triphenylmethyl radicals was not immediately accepted universally, since at that time carbon was believed to be invariably tetravalent in organic compounds.

The establishment of organic reaction mechanisms

Various types of evidence can be used to ascertain the probable mechanism of an organic reaction: these form the subjects of the remaining chapters of this book. Not all the methods necessarily apply to any one reaction, and the order of experimentation need not follow the order shown. The stage at which a hypothesis about a mechanism is advanced will vary. Sometimes this will have been before the original experiment, sometimes immediately after, and before further investigation. Quite frequently, however, new evidence may force a revised postulate about the mechanism, which may in turn suggest further tests.

PROBLEMS

1.1. Comment on the following reaction schemes on the basis of the considerations set out in pages 4 to 10. (Each of the processes indicated by an arrow should be considered as a possible elementary reaction).

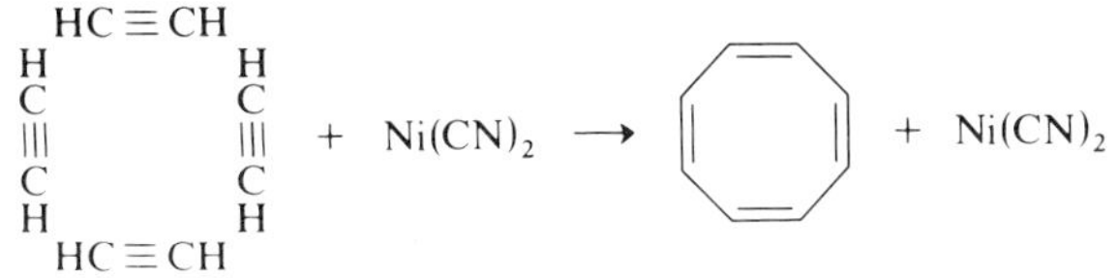
(a)

$$(CH_3)_4Pb \longrightarrow Pb + 4CH_3\cdot \qquad \text{(b)}$$

$$C_4H_9\cdot CH_2\cdot CH_2\cdot CH_2\cdot \overset{H}{C}HO\cdot \rightarrow C_4H_9\cdot CH_2\cdot CH_2\cdot \overset{H}{C}H\cdot \dot{C}H\cdot OH$$

$$\downarrow$$

$$C_4H_9\cdot CH_2\cdot \overset{H}{C}H\cdot \dot{C}H\cdot CH_2\cdot OH$$

$$\downarrow$$

products almost entirely from this radical $\leftarrow C_4H_9\cdot \dot{C}H\cdot CH_2\cdot CH_2\cdot CH_2\cdot OH \leftarrow C_4H_9\cdot \overset{H}{C}H\cdot \dot{C}H\cdot CH_2\cdot CH_2\cdot OH$

(c)

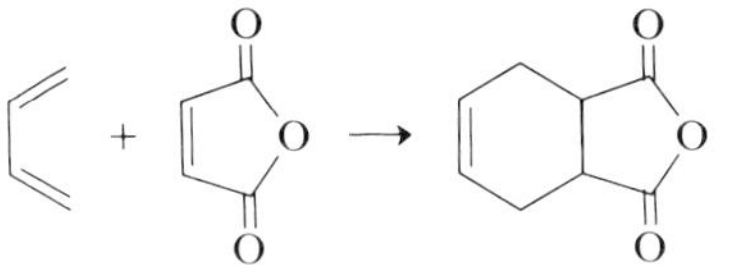
(d)

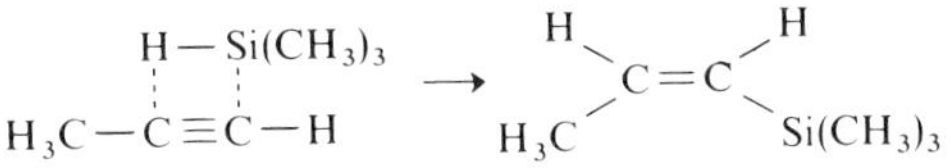
(e)

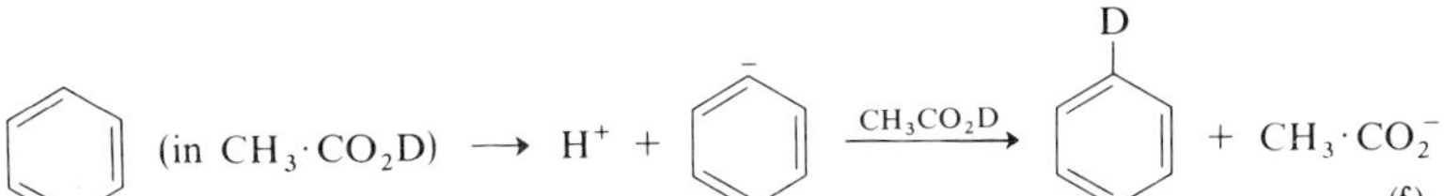
(f)

$$(CH_3)_3\overset{+}{N}{-}CH_2{-}CH_2{-}H \xrightarrow{\text{heat}} (CH_3)_3\dot{N}{-}CH_2{-}CH_2{-}H \rightarrow (CH_3)_3N + CH_2{=}CH_2$$

$$+OH^- \qquad\qquad +\cdot OH \qquad\qquad +H_2O \qquad \text{(g)}$$

2. Products

Establishment of the structure of products and their yields

I T M A Y seem banal, but the first step in investigating the mechanism of an organic reaction should be to establish qualitatively and then quantitatively the nature of the products. Without quantitative analysis of the products one cannot be certain that a given compound is the only, or even the major product of a reaction. Additionally, the identification of by-products can give important clues to the nature of the main reaction.

In a number of cases, mechanistic studies have been carried out on reactions whose products have been incorrectly identified. Triphenylmethyl, prepared from triphenylmethyl chloride by the action of zinc or other metals, exists in solution in equilibrium with a dimer. It gradually became accepted (see almost any textbook on physical organic chemistry published before 1968) that the dimer was hexaphenylethane, and on this basis a large number of studies of the equilibrium constant and the effect of substituents on it were carried out. Some of the results may have seemed surprising, for example the effects of *meta* and *para* methyl substituents (Table 1). There appears to be little difference in the degree of dissociation when a second methyl group is present in

TABLE 1

Apparent degree of dissociation (per cent) of some methyl-substituted 'hexaphenylethanes'

0·1 M solution in benzene (Tol = $CH_3C_6H_4$)

	'$(Ph_2TolC-)_2$'	'$(PhTol_2C-)_2$'	'$(Tol_3C-)_2$'
meta-Tol	6·5	7	40
para-Tol	5	5·5	16

the triarylmethyl radical, but the effect of introducing the third group is very marked indeed. Explanations based on the resonance effects of the methyl group do not adequately explain this jump, nor do they explain the more pronounced effect of *meta* compared with *para* substitution.

In 1968 it was demonstrated conclusively that the triphenylmethyl dimer has the structure **14**, not **13**. Hexaphenylethane would be expected to show only aromatic protons in its n.m.r. spectrum, whereas **14** would show olefinic protons additionally, as is observed. Thus the dimerization involves the attack of the α-carbon atom of one radical at a *para* position of another triphenylmethyl radical.

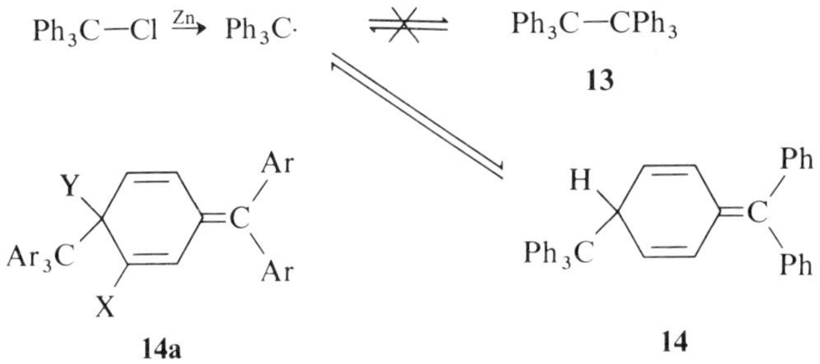

(25)

The revised formulation of the dimer as **14** allows more sense to be made out of the data in Table 1, since dimerization will be sterically hindered by both *meta* substituents (X in **14a**) and *para* substituents (Y in **14a**). The large effect of the third substituent is explained, since for radicals with only one or two substituents an unhindered position for *para* attack can be found, but for the tritolylmethyl radicals, no such unhindered *para* position exists.

Another example of the effects of incorrect structural assignments can be taken from the work of Thorpe, Ingold, and their collaborators in the 1920s. They were interested in the conditions which would favour the closure of cyclopropane rings, and proposed that bulky substituents on the β-carbon atom in compounds with the carbon skeleton **15** would repel each other, increase the $X-\hat{C}-X$ bond angle, and thereby decrease the $C-\hat{C}-C$ bond angle θ, making cyclopropane ring formation easier.

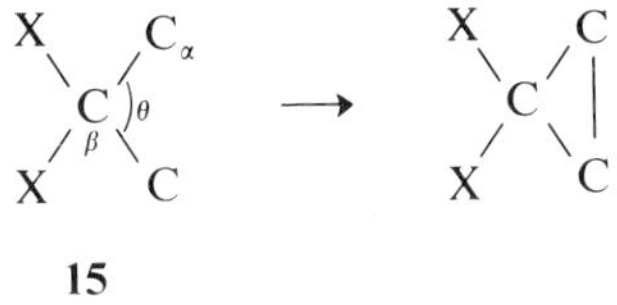

(26)

One of the reactions studied was the ring-closure reaction (27) which was postulated to be a reversible process, with the interconversion catalysed by strong base.

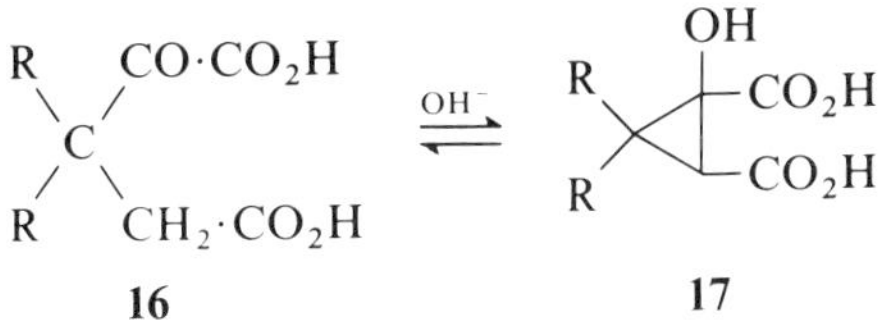

(27)

Experiments with different R substituents supported the above hypothesis. With R = Me, the equilibrium lay almost exclusively to the left, but when the two R groups were replaced by a spirocyclohexyl group the cyclopropyl compound **18** was favoured rather than the open chain form. At that time the

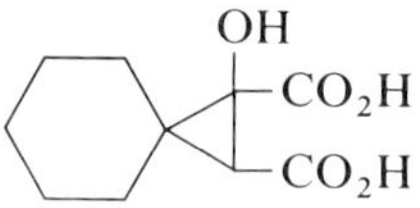

$$\text{cyclohexyl-spiro}\begin{cases}\text{OH}\\\text{—CO}_2\text{H}\\\text{—CO}_2\text{H}\end{cases}$$

18

cyclohexane ring was believed to be planar with bond angles of 120°, which on this theory should tend to contract the C—Ĉ—C angle θ. When R = ethyl, which should be intermediate between the cyclohexyl and dimethyl cases, an equilibrium was set up, in which the open chain (**16**) and cyclopropane (**17**) compounds were present in the proportions 38:62.

A re-investigation of the two compounds **16** and **17** (R = ethyl) by Wiberg and Holmquist in 1959, using the physical techniques of i.r. and n.m.r. spectroscopy revealed that neither **16** nor **17** had been assigned the correct structure. The compound formulated as **17** had no hydroxyl group (i.r.) and **16** was not a ketone (no reaction with 2,4-dinitrophenylhydrazine). The correct structures of **16** and **17** were shown by n.m.r. to be **19** and **20** respectively. It should be

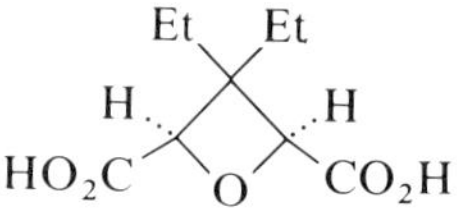

                19 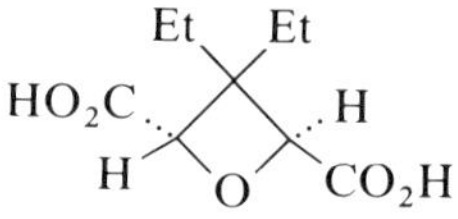   **20**

remembered that infrared spectroscopy had not been invented at the time of the original work, nor had 2,4-dinitrophenylhydrazine been established as a reagent for testing for ketones.

A stereochemical problem faced early investigators of the Beckmann rearrangement of oximes to amides. Unsymmetrical ketones RR′CO give two geometrically isomeric oximes RR′C=NOH, **21** and **22**, one of which rearranges on treatment with phosphorus pentachloride or sulphuric acid to

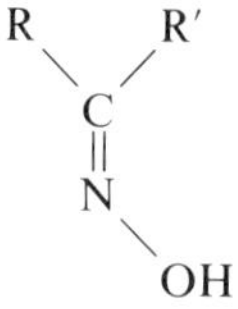

                21 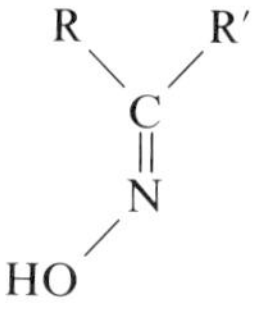   **22**

give the amide RCO·NHR′, the other to give R′CO·NHR. But which oxime is which? Hantzsch and Werner assigned the structures of the oximes by assuming that the nearer group would migrate.

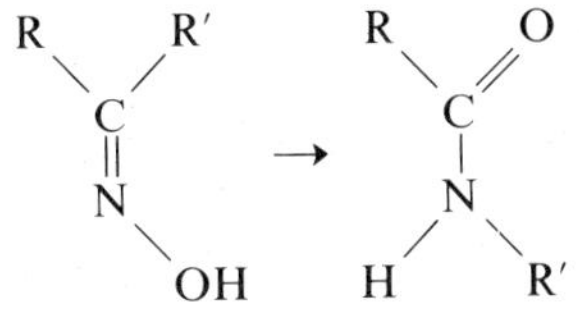

They were wrong. Before the advent of X-ray structures, chemical methods of assignment of structure had to be made, and were difficult to perform for the oximes. However, Meisenheimer and others were able to provide a number of separate arguments in favour of the opposite assignments of stereochemistry and thus *trans* migration of the R group in the Beckmann rearrangement. Two oximes of 2-bromo-5-nitroacetophenone can be prepared. One of these but not the other readily undergoes ring closure to the heterocyclic compound **24**, and therefore can reasonably be assigned the structure **23** (reaction 28).

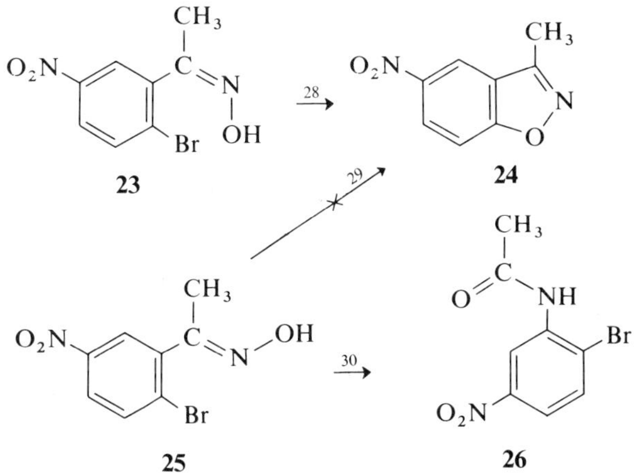

The isomer which does not undergo ring closure—**25**—undergoes the Beckmann rearrangement to give the amide **26**. Thus the rearrangement of **25** involves a *trans* migration of the phenyl group, and this conclusion was supported by a variety of other work.

Which bonds have been formed and broken during the reaction?

Once the structures of the starting materials and products have been established, the next thing is to find out which bonds have been made and broken during the reaction. This often appears to be obvious, but many examples of the obvious answer being wrong could be cited. For example, in nucleophilic substitution it may be assumed that the incoming nucleophile displaces the leaving group and becomes attached to the same carbon atom which formerly accommodated the leaving group.

$$HO^- + CH_3 \cdot CH_2 - Cl \rightarrow \overset{\beta}{CH_3} \cdot \overset{\alpha}{CH_2} \cdot OH + Cl^- \tag{31}$$

not

$$HO^- + \overset{H}{\underset{|}{CH_2}} - CH_2 - Cl \rightarrow HO - CH_2 - CH_2 - H + Cl^- \tag{32}$$

This assumption appears to be justified in most simple cases of this sort, but falls down in certain reactions which superficially appear to be analogous. For example the reaction of aromatic halides with potassium amide in liquid ammonia is thought to be a two-stage reaction (33) involving a benzyne intermediate (27), on the basis of experiments with ^{14}C-labelled iodobenzene (see also Chapter 1, p. 5).

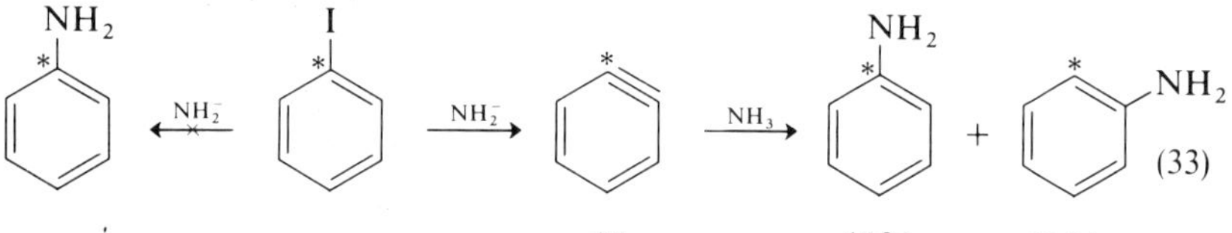

27   47%   53%   (33)

Isotopic studies are almost invariably the most expensive and time-consuming method of establishing which bonds have been broken or formed in a chemical reaction. However, the results are usually definitive, and often give information which could not be obtained by other methods. For example, when an ester is hydrolysed, does the oxygen atom from the water end up in the acid or in the alcohol molecule? In other words, does ester hydrolysis involve breakage of the alkyl–oxygen bond (reaction 34) or the acyl–oxygen bond (reaction 35)?

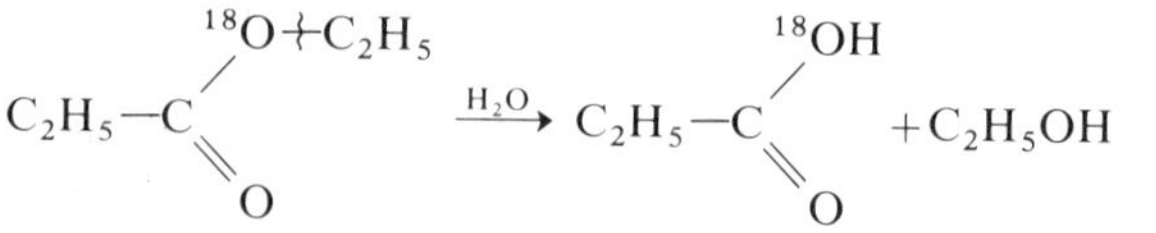

28

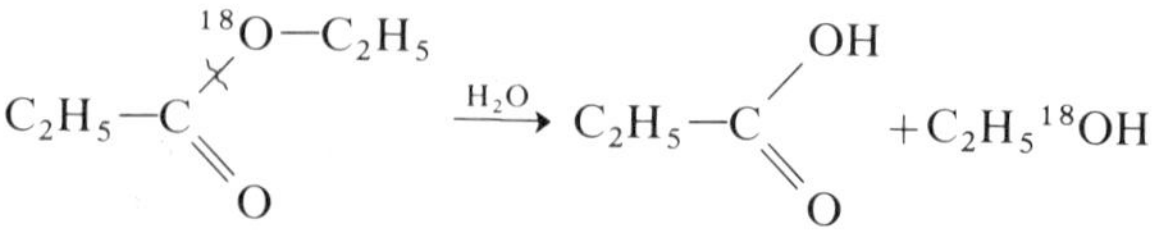

28

This question was resolved by using ethyl propionate **28** in which the oxygen marked ^{18}O was enriched with 0·7 per cent of ^{18}O. If alkyl–oxygen fission takes place (reaction 34) all the ^{18}O should remain in the acid, whereas if acyl–oxygen fission occurs (reaction 35) the ^{18}O should appear exclusively in the alcohol molecule. In fact, the ^{18}O appears exclusively in the alcohol, showing that under these experimental conditions acyl–oxygen bond breaking takes place (reaction 35).

Have groups or atoms moved from molecule to molecule during the reaction?

Many reactions involve the apparent migration of atoms or groups from one part of the molecule to another. Does the migrating group actually become free, or does it at all stages during the reaction remain attached to the original molecule? For example the allyl phenyl ether **29** rearranges on heating to the phenol **30**. Does the $-CH_2-CH=CH-C_2H_5$ group become free from the rest of the molecule during the reaction? This question may be

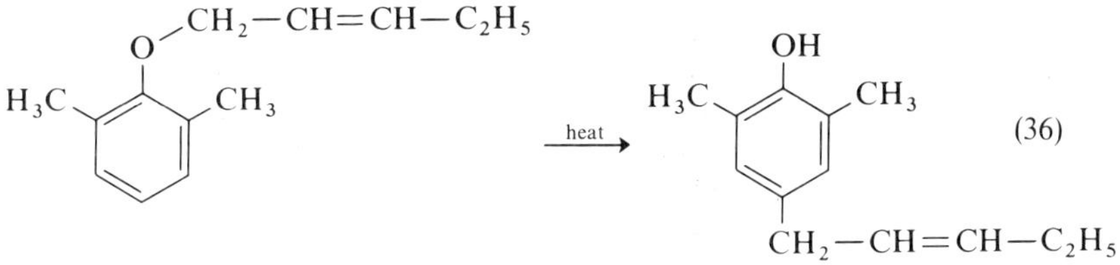

(36)

29   **30**

answered by designing a 'crossover' experiment, in which a similar molecule is synthesized, which contains small differences in both the 'migrating' and the 'stationary' parts of the molecule. Such a molecule could be **31**. After

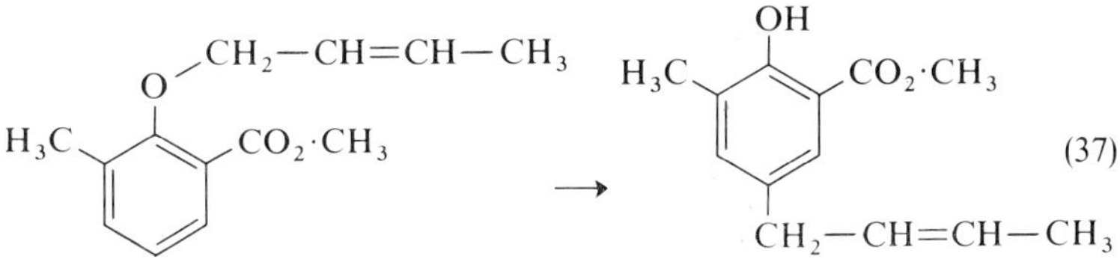

(37)

31   **32**

checking that the new compound **31** rearranges to **32** under the same conditions as the original ether at approximately the same rate, a mixture of the two ethers **29** and **31** is rearranged, and the resultant mixture of products is analysed to see if either of the crossover products **33** or **34** has been formed.

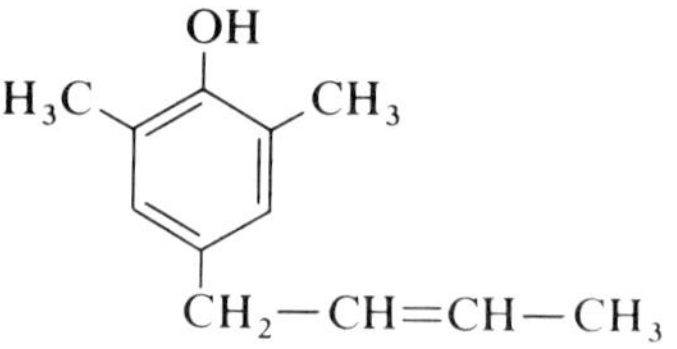

In this example, no trace of **33** or of **34** is detected, so it is safe to infer that the migrating group never becomes free of the rest of the molecule. Reactions of this type which occur 'within' a molecule are called intramolecular, as opposed to intermolecular reactions, in which groups are transferred from one molecule to another.

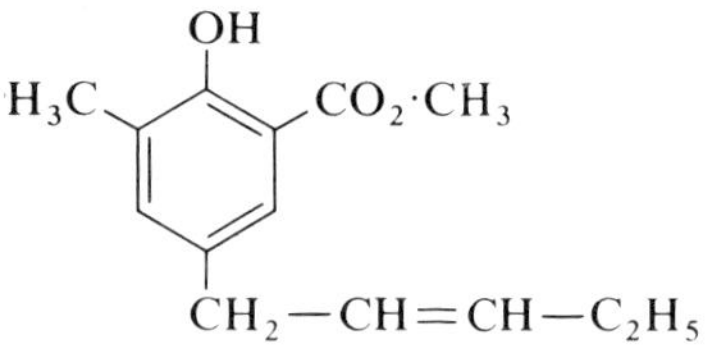

Lithium salts of benzyl ethers of the type **35** rearrange rapidly in inert solvents to give the alkoxides **36**. Is this reaction inter- or intra-molecular? A mixture of the two lithium salts **39** and **40**, which rearrange at similar rates, was heated. The products of the reaction indicated that although the reaction is predominantly intramolecular, about 7 per cent crossing-over of groups between molecules had occurred. This was taken as evidence that an 'intimate ion pair' **37** had been formed. The two ions would be held together in a cage of solvent molecules and would tend to react with each other to give the intramolecular product, although a small percentage of the ions could diffuse apart and combine at random. A mechanism involving a radical pair **38** rather than the ion pair would also fit these results, but the evidence accords neither with a complete dissociation of the ether into ions or radicals, nor with a direct intramolecular rearrangement in which the new carbon–carbon bond is formed as the carbon–oxygen bond breaks.

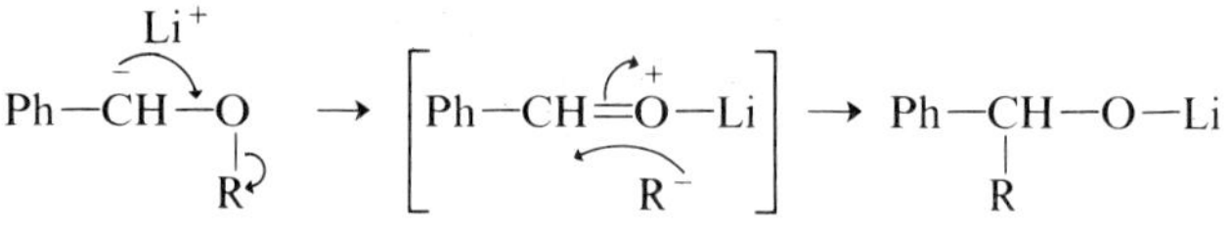

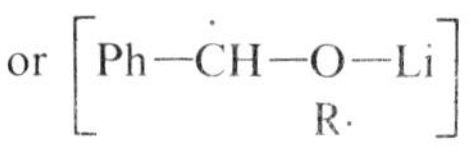

$$p-DC_6H_4\cdot CH(Li)-O-CHMeEt \qquad\qquad C_6H_5\cdot CH(Li)-O-CHMe_2$$
$$\textbf{39} \qquad\qquad\qquad\qquad\qquad \textbf{40}$$

Crossover experiments are only valid if the rates of reaction of the two compounds are approximately the same. If one compound rearranged (say) a hundred times faster than the other, then its rearrangement would be virtually complete before the reaction of the other had started, and thus even if the reaction were intermolecular, no crossover would be noted. A similarity in rate of rearrangement can most easily be achieved by using an isotopic label, but chemically labelled compounds of the type illustrated in reactions (36) and (37) are easier (and cheaper) to make and often pose easier analytical problems.

Exchange with solvent

Particularly in reactions involving movement of hydrogen from one molecule to another, there is a possibility that exchange of hydrogen with hydrogen atoms in the solvent may occur. The proposed mechanism should account for such occurrence or non-occurrence. For example two reactions which occur in water in the presence of bases are the self oxidation/reduction of formaldehyde (reaction 39) and the racemization of phenyl s-butyl ketone (reaction 41). In the first of these reactions it can be shown by carrying out the reaction in D_2O that no deuteration of the methyl group in the methanol takes place, and so it is a hydrogen atom from another molecule of formaldehyde which becomes attached. Hence we can rule out mechanisms such as (40) which involve hydride transfer from the hydroxyl group, and must postulate a mechanism such as (39) in which direct attack of one formaldehyde molecule on another takes place, with migration of carbon-bound hydrogen.

$$CH_2O \xrightleftharpoons{OH^-} \quad \overset{HO}{\underset{^-O}{>}}C\overset{H}{\underset{H}{<}} \cdots CH_2{=}\overset{\frown}{O} \rightarrow \overset{HO}{\underset{O}{>}}C\overset{H}{\underset{H}{<}} + \overset{H}{\underset{}{>}}CH_2{-}O^- \qquad (39)$$

$$CH_2O \xrightleftharpoons{OH^-} \quad \overset{H}{\underset{^-O}{>}}C\overset{O{-}H}{\underset{H}{<}} \cdots CH_2{=}\overset{\frown}{O} \rightarrow \overset{H}{\underset{HO}{>}}C{=}\overset{O}{} + \overset{H}{\underset{}{>}}CH_2{-}O^- \qquad (40)$$

On the other hand, when optically active phenyl s-butyl ketone **41** is racemized in aqueous dioxan in the presence of base, the racemization can be shown to be accompanied by exchange of hydrogen with the solvent by carrying out the reaction in D_2O. In fact, the rate of racemization is equal to the rate of deuterium incorporation, in support of the mechanism shown in (41).

$$Ph{\cdot}CO{\cdot}\overset{*}{C}HMeEt \xrightarrow{OD^-} Ph{\cdot}CO{\cdot}\overset{-}{C}MeEt \xrightarrow{D_2O} Ph{\cdot}CO{\cdot}CDMeEt \qquad (41)$$

41

optically active   racemic

Evidence from by-products and mixtures of products

Reactions rarely give 100 per cent of one product: usually other products are formed to a greater or lesser extent. It is possible that by-products may arise from a competing but slower reaction, but frequently they represent alternative ways in which reaction intermediates may react, and this gives information about likely reaction paths.

Methane can be fluorinated by molecular fluorine to give a mixture of CH_3F, CH_2F_2, CHF_3 and CF_4. Two possible mechanisms for fluorination would be the molecular process (42) or the free-radical scheme (43).

$$
\left.
\begin{array}{l}
\underset{\overset{|}{F}\underset{}{\rule[1mm]{10mm}{0.1pt}}\overset{|}{F}}{CH_3-H} \;\rightarrow\; \underset{F}{CH_3} + \underset{F}{H} \qquad
\underset{\overset{|}{F}\underset{}{\rule[1mm]{10mm}{0.1pt}}\overset{|}{F}}{CH_2F-H} \;\rightarrow\; \underset{F}{CH_2F} + \underset{F}{H} \\[4mm]
\underset{\overset{|}{F}\underset{}{\rule[1mm]{10mm}{0.1pt}}\overset{|}{F}}{CHF_2-H} \;\rightarrow\; \underset{F}{CHF_2} + \underset{F}{H} \qquad
\underset{\overset{|}{F}\underset{}{\rule[1mm]{10mm}{0.1pt}}\overset{|}{F}}{CF_3-H} \;\rightarrow\; \underset{F}{CF_3-H} + \underset{F}{}
\end{array}
\right\} \quad (42)
$$

Initiating reaction $\rightarrow F\cdot$

$$
\left.
\begin{array}{l}
F\cdot + CH_4 \;\rightarrow\; HF + CH_3\cdot \;\xrightarrow{F_2}\; CH_3F + F\cdot \\
F\cdot + CH_3F \;\rightarrow\; HF + CH_2F\cdot \;\xrightarrow{F_2}\; CH_2F_2 + F\cdot \\
F\cdot + CH_2F_2 \;\rightarrow\; HF + CHF_2\cdot \;\xrightarrow{F_2}\; CHF_3 + F\cdot \\
F\cdot + CHF_3 \;\rightarrow\; HF + CF_3\cdot \;\xrightarrow{F_2}\; CF_4 + F\cdot
\end{array}
\right\} \quad (43)
$$

However, C_2F_6 is always formed as a by-product in this reaction. This would be a natural consequence of the radical sequence: if $CF_3\cdot$ intermediates exist they are likely to dimerize to C_2F_6 as well as reacting with F_2 to give CF_4. Hence the presence of C_2F_6 among the products is evidence in support of the free-radical mechanism (43) rather than the molecular mechanism (42). (See also p. 10.)

The proposed mechanism should also account in a natural way for the presence of two or more products formed in substantial amounts, and for the non-formation of any products which might reasonably be expected, but which are absent.

Hydrolysis of the allylic chloride $Me_2C{=}CH{-}CH_2Cl$ by moist silver oxide gives a mixture of 15 per cent of the primary alcohol $Me_2C{=}CH{-}CH_2OH$ and 85 per cent of the isomeric alcohol $Me_2C(OH){-}CH{=}CH_2$. The same mixture of products, in almost exactly the same proportions, is obtained if $Me_2C(Cl){-}CH{=}CH_2$ is hydrolysed. Since this mixture is not the thermodynamic equilibrium mixture of the two isomers (which would contain about 90 per cent of the primary alcohol), this suggests that both halides react to give a common intermediate which then gives rise to the products by attack at two different positions. The most plausible intermediate is the mesomeric cation **42**.

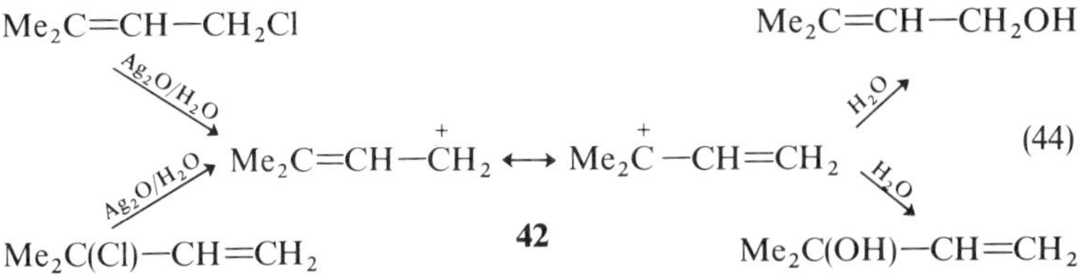

$$Me_2C=CH-CH_2Cl \qquad\qquad Me_2C=CH-CH_2OH$$

$$\xrightarrow{Ag_2O/H_2O} \quad Me_2\overset{+}{C}=CH-CH_2 \longleftrightarrow Me_2\overset{+}{C}-CH=CH_2 \qquad (44)$$

$$Me_2C(Cl)-CH=CH_2 \qquad\qquad \mathbf{42} \qquad\qquad Me_2C(OH)-CH=CH_2$$

The bromide $Me_2C=CH-CH_2Br$ when treated similarly gives the same mixture of products which supports the above conclusion.

Reactions involving aromatic compounds frequently involve the possibility of formation of two or more isomers. N-chloroacetanilide rearranges on treatment with HCl in chloroform to a mixture of o- and p-chloroacetanilide (69·8 : 30·2). Within experimental error, the same proportions of isomers (68·8 : 31·2) are formed when acetanilide is chlorinated directly in the same solvent. This suggests that the reaction *may* occur in two stages, the first involving production of chlorine, the second being the normal aromatic substitution of acetanilide, as shown in scheme (45). (This mechanism can be tested in other ways; for example, by crossover experiments and by trapping the chlorine intermediate). Conversely, the apparently analogous

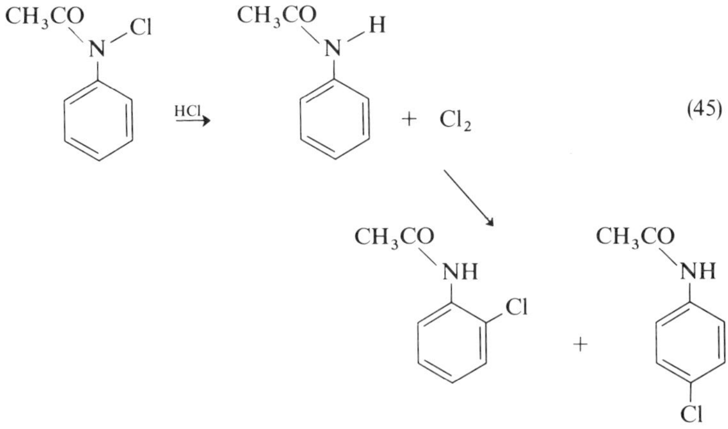

$$(45)$$

N-nitroaniline rearranges on heating with sulphuric acid to give mainly o-nitroaniline ($o:m:p = 93:0:7$). If free nitronium ions or other nitrating agents had been formed as intermediates, a mixture consisting mainly of *meta* and *para* isomers would be expected ($o:m:p = 6:34:59$ when aniline is nitrated with nitric acid under the same reaction conditions). This is therefore strong evidence that such an intermediate has not been formed, and that the nitro-group never becomes free of the remainder of the molecule, a conclusion confirmed (with some reservations) by other experiments.

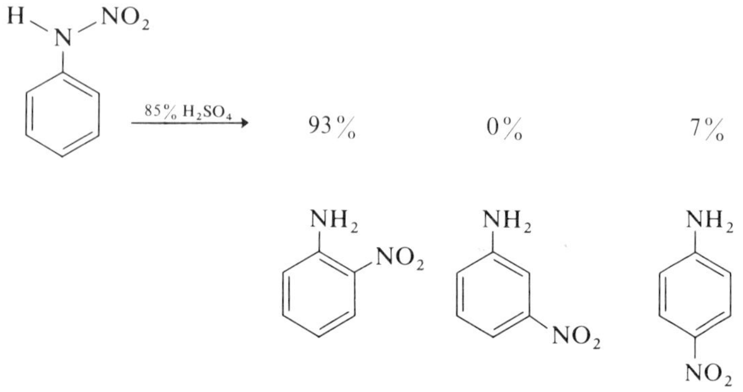

$$\text{(Ph)N(H)(NO_2)} \xrightarrow{\;85\% \text{ H}_2\text{SO}_4\;} \quad 93\% \qquad 0\% \qquad 7\% \quad (46)$$

$$\xrightarrow{\;\text{HNO}_3\;}_{85\% \text{ H}_2\text{SO}_4} \quad 6\% \qquad 34\% \qquad 59\% \quad (47)$$

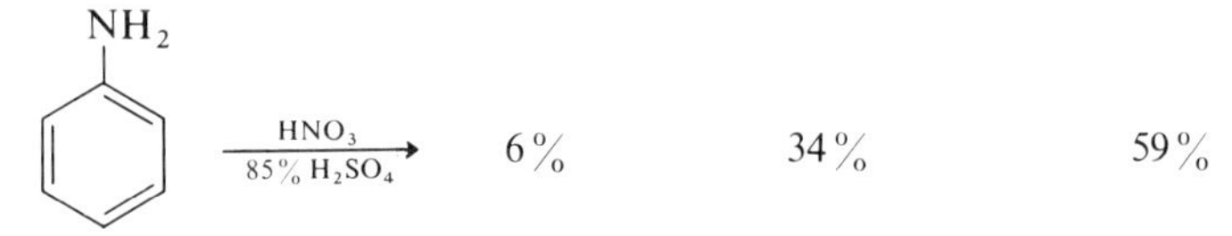

Summary

These examples show the importance of establishing the structures of the products and finding the yields of each. Minor products, by-products, and the absence of expected products can often suggest or rule out particular mechanisms. At the next level of refinement it may be determined which bonds have been broken and formed, whether the reaction is inter- or intra-molecular, and whether exchange of atoms with the solvent has taken place.

At this stage it is often possible (and helpful) to postulate one or several alternative mechanisms. These can be tested by the techniques described in the following chapters.

PROBLEMS

2.1. How would you show that simple nucleophilic substitutions do in fact follow mechanism (31) rather than (32) (two different experiments).

2.2. Could the results of reaction (5a) be explained by either of the following sequences.

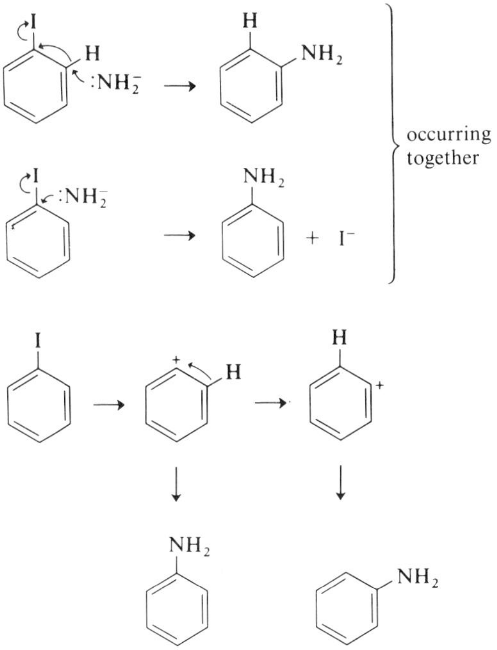

occurring
together

(a)

(b)

2.3. Suggest an experiment to prove that chlorine is formed as an intermediate in the acid-catalysed rearrangement of N-chloroacetanilide to o- and p-chloroacetanilide (reaction 45).

2.4. In reaction (39) how would you show that deuterium had not become attached to carbon in the methanol produced?

2.5. Benzaldehyde phenylhydrazone **A** reacts with oxygen to give a compound which has been formulated either as **B1** or as **B2**. The methyl substituted compound **C** does not react with oxygen. What information does this provide about the structure of **B**? What other evidence for the structure of **B** could be obtained?

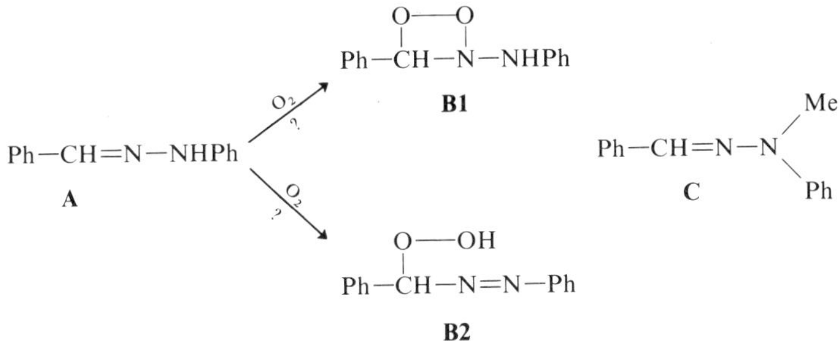

2.6. On treatment of R_3C-CH_2Cl with base, a substantial yield of the olefin $R_2C=CHR$ is produced. The rearrangement, which is very much faster for $R = Ph$ than for $R = CH_3$ is thought to take place by the mechanism:

$$R_3C-CH_2-Cl \rightarrow R_2\overset{R}{\underset{|}{C}}\overset{+}{C}H_2 \rightarrow R_2\overset{+}{C}-CH_2R \xrightarrow{-H^+} R_2C=CHR$$

If a crossover experiment was performed with a mixture of Ph_3C-CH_2Cl and Me_3C-CH_2Cl, and no $Me_2C=CHPh$ or $Ph_2C=CHMe$ was produced, would this be good evidence that R groups never become free during the reaction?

3. Kinetics

AFTER the establishment of the reaction products, probably the most important feature of a reaction is the rate at which it takes place. The insight into mechanism given by a study of rate forms the subject of this chapter.

Experimental technique

Ideally, we should like to determine the concentrations of reactants and products (under particular conditions) as a function of time after mixing. Any chemical or physical method of analysis appropriate for a particular reactant or product may be used. Methods of analysis fall into two groups: those in which aliquots are taken at different times during the reaction (either by withdrawing a sample from the reaction mixture, or by sealing the reaction mixture into a number of ampoules which are withdrawn and opened at intervals), and those in which a physical feature of the system may be monitored continuously without disturbing the reaction.

The first group of techniques includes titration (for acids, bases, oxidizing and reducing agents, halide ions, etc.), weighing of precipitates, and separation of products by gas chromatography. These techniques are direct; they can be very accurate and tend to be specific for particular compounds with little danger of misinterpretation of results. However, the process of sampling is often inconvenient or time-consuming, and these difficulties are avoided by the second group of analytical methods, which are physical methods that do not disturb the reaction vessel. Such techniques include dilatometry (measurement of volume changes of the reaction mixture), conductivity measurements, electrode potential measurements, and the various forms of spectroscopy, particularly ultraviolet and, to a lesser extent, infrared and n.m.r. These methods are often convenient, and allow a large number of readings to be made during the course of the reaction. However, the results are more dependent on interpretation. If, for example, a reaction is being followed by the build-up of absorption at a particular wavelength at which one of the products is known to absorb, then it must be established that the observed absorption is not due to a small amount of a different compound which happens to have a large extinction coefficient at this wavelength.

Ideally, several different techniques should be used to follow a reaction.

Fast reactions

Normal techniques of chemical analysis are convenient for reactions with half-lives greater than about five minutes. Physical methods bring this down to about ten seconds without any elaborate apparatus. For reactions faster than this, flow, pulse or relaxation techniques may be used: these methods allow reactions with half-lives down to about 10^{-9} second to be studied in favourable cases.

Over what range of concentrations and times should the reactions be studied?

For a particular reaction, this should normally be as large as possible, say at least to 80–95 per cent completion of reaction. If a physical method such as u.v. absorption spectroscopy is being used to follow the reaction there is usually no problem in obtaining a large number of readings spread through the reaction time. However when aliquots are taken, and particularly when the reaction mixture is sealed in ampoules, only a limited number of readings can be taken, and a decision must be made as to when to take them. In a typical ampoule experiment, there might be only be ten ampoules. It is usually best to spread the readings out evenly over two or three halflives, leaving two ampoules for 'infinity' readings if these are necessary.

In different experiments, we should aim at covering the maximum possible range of concentrations (often this is limited by solubility, reaction times, etc.), and if a temperature study is being made, the widest possible range of temperatures should be used. Much more useful information will be gathered by such efforts than by concentration on a very large number of repetitions of experiments under one set of conditions. Duplication (at least) of results should however be routine.

Precautions

The rates of many reactions are profoundly altered by the presence of air in the reaction system or by the influence of light. Air can be excluded by carrying out the reactions under argon or nitrogen, or still more rigorously by sealing the reaction mixtures into glass ampoules, after 'de-gassing' them on a vacuum line. Light is simply excluded by painting the outside of the reaction vessels black (or by wrapping them in metal foil), or by carrying out the experiments in a dark-room.

It goes without saying that the reagents and solvents used should be of the highest possible purity to avoid uncertainties due to contaminants. A classical example of difficulties caused by impurities, as well as the presence or absence of air or light is the addition of hydrogen bromide to terminal olefins, mentioned briefly in Chapter 1, and considered in more detail in Chapter 4. In this reaction, mixtures of the corresponding 1-bromoalkane and 2-bromoalkane are produced in proportions which tend to vary from experiment to experiment. However, when the reagents are carefully purified and the reaction is carried out in the absence of oxygen and light, the product is almost exclusively the 2-bromoalkane. Under these conditions, the reaction leading to the 2-bromoalkane can be studied in isolation from the other reaction which gives the 1-bromoalkane.

Some reactions proceed at surfaces, rather than homogeneously through-out the liquid or gas. This problem is greater for gas-phase than for liquid-phase reactions. A check that surface reactions are unimportant can be made by greatly increasing (say by a factor of ten) the surface of the vessel, which may be achieved for a glass vessel by filling it with glass rods, glass beads, or glass wool. If no change in the rate is observed when a major change in the surface/volume ratio of the vessel is made, it is concluded that the reaction is homogen-ous and does not take place at the surface. If reaction does take place at the surface, it may be possible to prevent surface reactions by changing the material of the vessel, or by coating it with an inactive material.

Rates of reaction

What is usually measured is the concentration of one or more reaction components after different times of reaction. More useful from the point of view of interpretation is the rate of the reaction, i.e. dc/dt, for a particular starting material or product. The rate of a reaction at any particular time is normally obtained by plotting a concentration-versus-time graph, drawing a smooth curve through the experimental points, and then drawing the tangent to this graph at different points.

This process is rather tedious and is liable to error. In some cases the need to obtain rates in this way can be avoided by making assumptions about the order of reaction (see below).

Order of reaction

For a very large number of reactions the rate of reaction is found to vary with the concentrations of one or more of the reactants according to an expression of the type

$$\frac{-d[A]}{dt} = k[A]^l[B]^m[C]^n... \tag{48}$$

where A, B, C, ... are reactant molecules. The powers $l, m, n, ...$ may or may not be whole numbers. If this expression is obeyed, the reaction is said to be of total order $(l+m+n...)$ and lth order with respect to A, mth order with respect to B, nth order with respect to C and so on. In practice, rarely more than two reactants are involved in such an expression, and quite frequently only one. The constant k is the $(l+m+n+...)$th order rate constant for the reaction.

One way of evaluating the orders of reaction with respect to each com-ponent is to measure the rates of reaction in experiments in which all com-ponents except one are present in large excess, so that their concentrations may be regarded as constant. The logarithm of the rate of reaction (obtained as above) is then plotted against the logarithm of [A] the concentration of the remaining reactant A.

$$\log(\text{rate}) = \log(k') + l \log[A].$$

The slope gives l, the order of reaction with respect to A, and the intercept gives k' (which includes the concentration terms involving B, C, etc.) The experiments are then repeated with A and C in a large excess to determine the order with respect to B, and so on. Hence the orders of reaction with respect to each component can be deduced, and likewise the rate constant k.

This method of determining orders, and the need to deduce rates from the raw data is often bypassed in favour of the following method. A rate law for the reaction is assumed (often by analogy with other reactions, or by an assumption about the mechanism). This expression is integrated to give the concentration of the reagents as a function of time. For example in a reaction with only one component, where it is assumed that the reaction is of second order, i.e. $-d[A]/dt = k_2[A]^2$, this expression can readily be integrated to:

$$\frac{1}{[A]} - \frac{1}{[A_0]} = k_2 t$$

Hence to test the hypothesis (that the reaction is second order) the reciprocal of the concentration of A should be plotted against time. If the graph is a straight line this is good evidence that the reaction is of second order, provided that the reaction has been carried out for at least two half-lives. In a similar way, a graph of $\log[A]$ against time can be used to test (by its straightness) for first-order behaviour. Similar plots may be constructed to cover other orders, including fractional orders.

The advantage of testing for different orders of reaction in the above way is that it avoids the need to take tangents from the original curve. Its disadvantage is that it usually fails to reveal small deviations from (assumed) integral orders, particularly if a wide range of concentrations has not been studied.

Reaction order from reverse reactions

Frequently the equilibrium constant for a reaction is so small that the rate of the forward reaction is difficult to measure. If the reaction is elementary (i.e. no intermediates), the law of mass action applies, and equation (49)

$$\frac{k_{forward}}{k_{reverse}} = K \tag{49}$$

holds even when the reaction is not at equilibrium. Hence if the equilibrium constant is known (measured or calculated), and the kinetic order and reaction rate constant is determined for the reverse reaction, the order and rate constant for the forward reaction can be calculated from (49). If the reaction is not elementary, relationship (49) will still hold and provide kinetic information about the forward reaction over the same range of concentrations as has been used for the reverse reaction, provided that

(a) the reaction is not a chain reaction (see p. 40)
(b) any intermediates formed are present in only very low concentrations
(c) a steady state (p. 39) is quickly set up.

An important example of the use of this method is provided by the base-catalyzed dimerization of acetone to diacetone alcohol (DAA), reaction (50), the mechanism of which is considered later in this chapter. The equilibrium constant for reaction (50) favours the reverse reaction (i.e. the production of acetone), and from dilatometric studies the rate of the reverse reaction is given as $k_{-50}[DAA][OH^-]$. Since the equilibrium constant can be determined,

$$2\ CH_3 \cdot CO \cdot CH_3 \underset{OH^-}{\overset{OH^-}{\rightleftharpoons}} CH_3 \cdot CO \cdot CH_2 \cdot C(OH)(CH_3)_2 \qquad (50)$$

$$DAA$$

$$K_{50} = k_{50}/k_{-50} = \frac{[DAA]}{[CH_3 \cdot CO \cdot CH_3]^2} \qquad (50a)$$

the kinetic order of the forward reaction may be found by casting it in the general form:

$$rate = k_{50}[CH_3 \cdot CO \cdot CH_3]^x[DAA]^y[OH^-]^z$$

and equating this at equilibrium with the rate of the reverse reaction, $k_{-50}[DAA][OH^-]$. Comparison of expression (50b) with (50a) shows that $x = 2$, $y = 0$, and $z = 1$, thus giving the kinetic order of the (not conveniently measurable) forward reaction.

$$\frac{k_{50}}{k_{-50}} = \frac{[DAA][OH^-]}{[CH_3 \cdot CO \cdot CH_3]^x[DAA]^y[OH^-]^z} \qquad (50b)$$

The connection between mechanism and kinetics

A proposed mechanism requires particular kinetic behaviour, but the reverse is not true. A reaction mechanism cannot be deduced unequivocally from the kinetics. However, kinetic studies may allow us to rule out particular mechanisms, or they may allow a choice to be made between alternative mechanisms which appear plausible for a reaction.

Reactions of integral order

Reaction rate theory establishes the kinetic behaviour for an elementary reaction: the order with respect to each component is equal to the number of molecules of that component taking part in the reaction. Only three basic types of elementary reaction are kinetically important (Table 2).

TABLE 2

Kinetically important reaction types†

	Total order	Order with respect to	
		A	B
A → Products (51)	1	1	0
A + A → Products (52)	2	2	0
A + B → Products (53)	2	1	1

† Termolecular reactions are ignored in this discussion.

It is useful to define the molecularity of an elementary reaction as the number of molecules involved as reactants. Thus reaction (51) is unimolecular, whereas (52) and (53) are bimolecular. For an elementary reaction the molecularity is the same as the kinetic order. It should be emphasized that kinetic order is an experimental feature of the reaction, whereas the molecularity is a feature of the individual elementary steps of a proposed mechanism.

If a reaction shows any of the three kinetic orders shown in Table 2, then this behaviour is consistent with the reaction being elementary, of type (51), (52), or (53) as appropriate. This is the simplest explanation of the results, though it may turn out from other evidence to be incorrect. Examples of elementary reactions of the three types are:

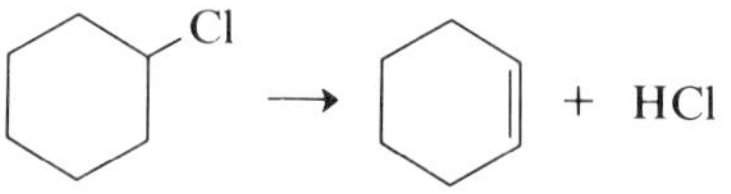

(54)

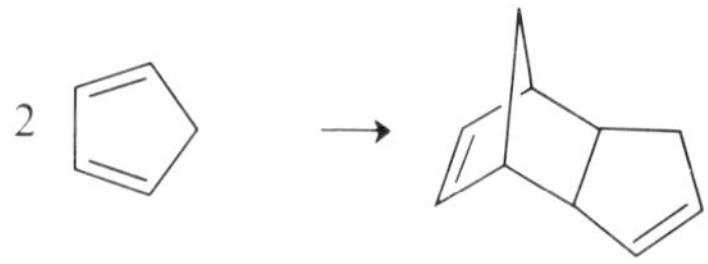

(55)

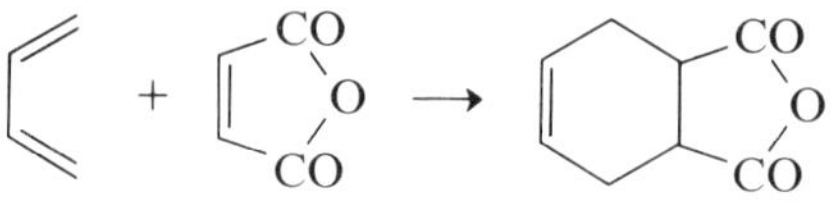

(56)

As an example of the use of kinetics to differentiate between different possible mechanisms, let us consider nucleophilic substitutions of organic halides. Two of the most plausible mechanisms are shown in (57) and (58). The symbol Nu:⁻ indicates a negatively charged nucleophile (p. 3).

$$S_N2 \quad Nu:^- + R-Hal \rightarrow Nu-R + Hal^- \tag{57}$$

$$S_N1 \quad R-Hal \underset{-58a}{\overset{58a\ (slow)}{\rightleftharpoons}} Hal^- + R^+ \xrightarrow[\text{fast} \quad 58b]{Nu:} Nu-R \tag{58}$$

Mechanism (57) would lead to second-order kinetics, first-order with respect to each component, as found for a number of reacting systems. Mechanism (58) is complex and consists of two steps. If it is assumed (as is likely) that (58a) will be slow, and also that (58b) is fast compared with the reverse of (58a) (not always true, see below), then the rate of the reaction will depend entirely on step (58a), since any R^+ formed will react rapidly with the nucleophile to give Nu—R. The observed kinetics will therefore be first order with respect to the alkyl halide and zero order with respect to the nucleophile (independent of the nucleophile concentration). Reaction (58) is an example of a reaction with a *rate-determining step*, since the kinetics depend only on reaction (58a). In reactions with a rate-determining step no kinetic information is available (without studying a different reaction system) about reaction steps after the rate-determining step. In reaction (58), the rate of (58b) could be increased or decreased by large factors without changing the overall rate of destruction of R—Hal.

Thus an experimental distinction can be made between mechanisms (57) (bimolecular, S_N2) and (58) (unimolecular, S_N1) for nucleophilic substitution reactions. Reaction scheme (58) for S_N1 reactions can be tested further in varous ways. Simple first-order kinetics will be observed if reaction (−58a) is very slow. However, if reaction (−58a) is comparable in rate with (58b), the rate of reaction will fall with time faster than is predicted from the first-order rate equation, since when a significant concentration of halide ion has been built up, alkyl halide will be reformed by reaction (−58a). A decrease in rate will also be observed if the corresponding halide ion is added to the reaction mixture at the start of the experiment. This reduction in rate is often referred to as the *common ion* or mass law effect. Reaction scheme (58) also suggests that if a mixture of nucleophiles were used rather than a single nucleophile, a mixture of products should be obtained but the rate of substitution should not be altered. When sodium azide is added to a solution of *p,p'*-dimethyl-benzhydryl chloride, **43**, in aqueous acetone, there is no significant change in rate (apart from a salt effect†), but the product becomes a mixture of the alcohol **44** and the azide **45**. The concentration ratio [**45**]/[**44**] increases linearly with

---

† Reactions in which neutral molecules give ions, or in which ions of the same sign react together, have rates which are increased in the presence of salts (i.e. ions). Conversely the rates of reactions between ions of opposite sign are decreased in the presence of salts. The salt effect may be used as a diagnostic test for mechanism: in the past it has been applied mainly to inorganic rather than to organic reactions.

azide concentration. Since the water concentration is effectively constant, this is what would be expected from a competition between reactions (59a) and (59b)

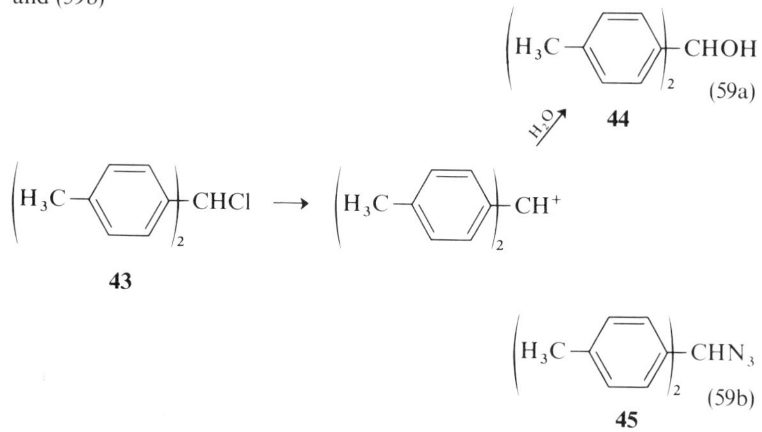

$$\left(H_3C \!\!-\!\!\left\langle \bigcirc \right\rangle\!\!-\!\! CHOH \right)_2 \quad (59a)$$

44

$$\left(H_3C \!\!-\!\!\left\langle \bigcirc \right\rangle\!\!-\!\! CHCl \right)_2 \;\longrightarrow\; \left(H_3C \!\!-\!\!\left\langle \bigcirc \right\rangle\!\!-\!\! CH^+ \right)_2$$

43

$$\left(H_3C \!\!-\!\!\left\langle \bigcirc \right\rangle\!\!-\!\! CHN_3 \right)_2 \quad (59b)$$

45

Different reactions which occur at the same rate

In a number of cases, several reactions of a particular compound are found to proceed at the same rate. This is unlikely to occur by coincidence, particularly if the similarity in reaction rate persists at more than one temperature. The most likely explanation is that all these reactions have a common mechanism up to and including the rate-determining step, and that reagents which give rise to the different products are involved only in subsequent stages. Examples of this type of behaviour include the base-catalysed chlorination, bromination and iodination of acetone, and the related reactions of phenyl s-butyl ketone, where in addition to halogenation reactions, deuterium exchange and racemization of the optically active ketone all proceed at the same rate. In each of these two groups of reactions, the reaction is first order with respect to the base and to the ketone, suggesting that the abstraction of a proton from the ketone is the rate-determining step, which is followed by fast reactions of the anion.

$$HO^- + Ph \cdot CO \cdot CHMeEt \xrightarrow{\text{slow}} Ph \cdot CO \cdot C^- MeEt \xrightarrow{H_2O} Ph \cdot CO \cdot CHMeEt$$
(racemized)

$$\xrightarrow{D_2O} Ph \cdot CO \cdot CDMeEt$$

$$\xrightarrow{Br_2} Ph \cdot CO \cdot CBrMeEt$$

(60)

Reactions of order higher than second

Since elementary reactions are normally unimolecular or bimolecular, a reaction which shows a kinetic order higher than second cannot be elementary. Nevertheless, many reactions are kinetically of third or even higher order.

This kinetic behaviour is often caused by a reaction mechanism which involves a number of stages with a reversible reaction early on in the process, and a rate-determining step later. A typical scheme would be:

$$A + B \rightleftharpoons C \tag{61}$$

$$C + D \rightarrow E \tag{62}$$

Provided that
(a) reaction (62) is slower than both (61) and its reverse, and
(b) the equilibrium for reaction (61) lies well to the left,
the equilibrium corresponding to reaction (61) will soon be set up, and the concentration of C will be given by:

$$[C] = (k_{61}/k_{-61})[A][B] \tag{63}$$

The rate of production of E, and hence the overall rate of reaction will be given by (64). Hence third-order kinetics will be observed.

$$\frac{d[E]}{dt} = -\frac{d[A]}{dt} = -\frac{d[B]}{dt} = k_{62}[C][D] = (k_{61}k_{62}/k_{-61})[A][B][D] \tag{64}$$

Reactions involving this type of behaviour are very common, and include a large number of reactions catalysed by acids or bases. An example of acid catalysis is esterification (65):

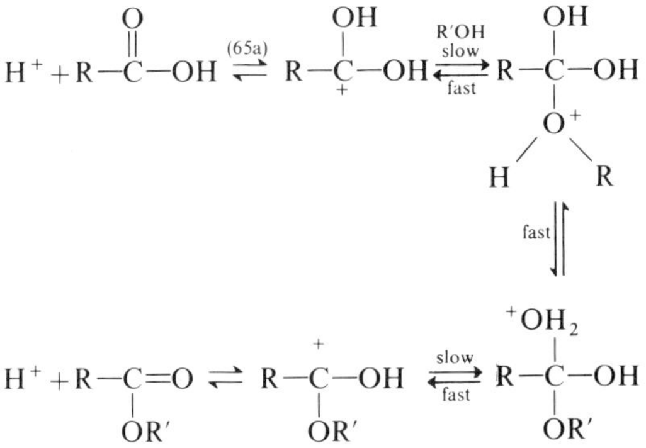

This scheme accounts for the observed dependence of the rate of the reaction on $[H^+][RCO_2H][R'OH]$.

Base-catalysed reactions which show third-order behaviour include the important aldol condensation of aldehydes or ketones, illustrated for acetone (66) and (67). [For this particular reaction, since the equilibrium lies well to the left, most kinetic work has been done on the reverse reaction, from which the kinetics of the forward reaction may be deduced (see page 29)].

$$CH_3 \cdot CO \cdot CH_3 + OH^- \overset{\text{fast}}{\rightleftarrows} H_2O + CH_3 \cdot CO \cdot CH_2^- \tag{66}$$

47

$$CH_3 \cdot CO \cdot CH_2^- + CH_3 \cdot CO \cdot CH_3 \overset{\text{slow}}{\rightleftarrows} CH_3 \cdot CO \cdot CH_2 - \underset{\underset{CH_3}{|}}{\overset{\overset{O^-}{|}}{C}} - CH_3$$

$$\overset{\text{fast}}{\rightleftarrows} CH_3 \cdot CO \cdot CH_2 \cdot \underset{\underset{CH_3}{|}}{\overset{\overset{OH}{|}}{C}} - CH_3 \tag{67}$$

Why do so many acid- and base-catalysed reactions follow third-order behaviour? Proton transfer reactions in general tend to be fast, and the equilibria of a large number of reactions of this type lie well over to the left because most protonated organic compounds (e.g. **46**) are more acidic than H_3O^+, and most organic bases formed by abstraction of a proton from a C—H bond (e.g. **47**) are stronger bases than OH^-, OEt^- etc. Hence the equilibria for reactions (65a) and (66), and many others like them lie well to the left, and subsequent reactions of the organic intermediate can give rise to third-order kinetic behaviour.

Complex reactions which show deceptively simple kinetics

In general, the first mechanistic hypothesis to be tried is the simplest. For a reaction with the overall stoichiometry $A + B \longrightarrow C + D$ which shows second-order kinetics, first order with respect to both A and B, the simplest hypothesis is that the reaction is bimolecular, and occurs between one molecule of A and one of B. A variety of evidence supports the view that, for example, the second-order nucleophilic substitution reactions (57) are in fact simple (elementary) bimolecular reactions.

However, simple kinetic behaviour is sometimes shown by reactions whose mechanism is complex. An inorganic example is the reaction between hydrogen and iodine. This reaction, which is reversible, follows second-order kinetics, first order with respect to hydrogen and to iodine. The 'obvious' mechanism, accepted for sixty years, was that this is a simple bimolecular reaction (68)

$$\begin{matrix} H-H \\ I-I \end{matrix} \longrightarrow \begin{matrix} H & H \\ | & | \\ I & I \end{matrix} \tag{68}$$

Indeed it would seem perverse without further evidence to suggest that the reaction should proceed by any other mechanism. This evidence was duly discovered in 1967 by Sullivan, who proved that the reaction involved iodine atoms which are present in thermal equilibrium with iodine molecules.

Let us consider an alternative mechanistic scheme [reactions (69) and (70)] for the hydrogen–iodine reaction, in which an equilibrium is set up between iodine molecules and iodine atoms [reaction (69)] followed by a slow reaction of the iodine atoms with a hydrogen molecule [written for simplicity as a single-step process, though reaction (70) may well take place in two steps].

$$I_2 \underset{\text{fast}}{\overset{\text{fast}}{\rightleftharpoons}} 2I\cdot \qquad (69)$$

$$2I\cdot + H_2 \rightleftharpoons 2HI \quad \text{(in one or two stages)} \qquad (70)$$

Since the equilibrium concentration of iodine atoms will be given by

$$[I\cdot] = ([I_2]\, k_{69}/k_{-69})^{\frac{1}{2}},$$

the rate of production of HI early in the reaction, when the back reaction (i.e. $2HI \longrightarrow H_2 + I_2$) can be neglected, will be given by

$$\frac{d[HI]}{dt} = 2[I\cdot]^2[H_2]k_{70} = \frac{2k_{69} \cdot k_{70}}{k_{-69}}[H_2][I_2]. \qquad (71)$$

Thus no distinction between the simple bimolecular mechanism (68) and the complex mechanism (69) and (70) can be made on the basis of the kinetics of the thermal reaction, and a distinction must be made in some other way. This was achieved by the photochemical production of iodine atoms in a mixture of iodine and hydrogen at temperatures low enough for the thermal reaction rate to be negligible. From the rate of production of hydrogen iodide, a value of k_{70} can be found. The ratio of rate constants k_{69}/k_{-69} is the known equilibrium constant for the dissociation of iodine molecules into atoms at a particular temperature. Hence Sullivan was able to calculate from equation (71) the rate at which hydrogen iodide would be produced in the thermal system by the mechanism (69) and (70). Within experimental error, this was the rate at which the thermal reaction did in fact take place, and hence the contribution from the direct bimolecular reaction (68) must be negligible.

Reactions which are not of integral order

Not all reactions show a simple dependence of rate on integral orders of one or more reaction components. Even for reactions of integral order we have seen that the reactions need not necessarily be elementary. Reactions which show more complex kinetic behaviour cannot be elementary. Two commonly occurring types of non-integral order behaviour are described below, with a discussion of the mechanistic information which can be obtained.

Mixed-order reactions

Many reactions show a mixed kinetic order in which the rate can be expressed as the sum of two or more terms, for example (72) and (73):

$$\text{Rate} = k_1[\text{A}] + k_2[\text{A}][\text{B}] \tag{72}$$

$$\text{Rate} = k_1[\text{A}][\text{B}] + k_2[\text{A}][\text{B}]^2 \tag{73}$$

Experimentally, it is often difficult to distinguish between this type of kinetic dependence and a fractional kinetic order dependence, e.g.:

$$\text{Rate} = k[\text{A}][\text{B}]^{0.5} \tag{74}$$

$$\text{Rate} = k[\text{A}][\text{B}]^{1.5} \tag{75}$$

unless a very considerable range of concentrations of B can be covered. The usual way of testing for such mixed-order behaviour can be shown for a particular example for which behaviour of type (72) is anticipated. If equation (72) is divided through by [A], an expression for the 'apparent first-order rate constant', Rate/[A] is given by (76).

$$\text{Apparent first order rate constant} = \text{Rate}/[\text{A}] = k_1 + k_2[\text{B}] \tag{76}$$

The apparent first-order rate constant is plotted against [B] in Fig. 1. If the reaction really follows kinetic behaviour of type (72) this graph will be a straight line of slope k_2 and intercept k_1. Similar plots for other types of mixed-order behaviour can readily be made.

The usual reason for mixed-order behaviour is that one of the components of the reaction mixture is reacting by two different routes. An example is the hydrolysis of isopropyl bromide by sodium hydroxide in aqueous ethanol, where the observed dependence of the rate on the reagent concentrations is given by:

$$\text{Rate} = \frac{-\text{d}[\text{Pr}^i\text{—Br}]}{\text{d}t} = k_1[\text{Pr}^i\text{—Br}] + k_2[\text{Pr}^i\text{—Br}][\text{OH}^-] \tag{77}$$

$$\underset{\substack{\text{S}_\text{N}1 \\ \text{component}}}{\phantom{k_1[\text{Pr}^i\text{—Br}]}} \quad \underset{\substack{\text{S}_\text{N}2 \\ \text{component}}}{\phantom{k_2[\text{Pr}^i\text{—Br}]}}$$

This is interpreted in terms of concurrent hydrolysis by the $\text{S}_\text{N}1$ and by the $\text{S}_\text{N}2$ mechanisms discussed earlier. Since under these reaction conditions ethyl bromide reacts exclusively by the $\text{S}_\text{N}2$ mechanism whereas t-butyl bromide reacts by the $\text{S}_\text{N}1$ mechanism, it is not surprising that both hydrolysis routes are possible for the intermediate isopropyl bromide (the reason for these differences in behaviour will be discussed in Chapter 6).

Many acid- and base-catalysed reactions show mixed-order kinetics. For example, ethyl ortho-acetate **48** is hydrolysed to ethanol and acetic

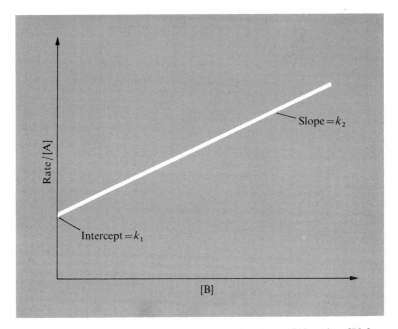

FIG. 1. Mixed first- and second-order kinetics. Plot of Rate/[A] against [B] for a reaction which follows the rate expression: rate $= k_1[A] + k_2[A][B]$.

acid (reaction 78) in aqueous buffers of m-nitrophenol (a weak acid) and its sodium salt. At constant ionic strength, the rate of disappearance of ethyl ortho-acetate **48** was shown to follow the kinetic expression (79).

$$CH_3-\underset{\underset{OEt}{|}}{\overset{\overset{OEt}{|}}{C}}-OEt \xrightarrow[\text{slow}]{\text{HA}} CH_3-\underset{\underset{OEt}{|}}{\overset{\overset{OEt}{|}}{C}}-\overset{+}{\underset{H}{\overset{Et}{\overset{/}{O}}}} \xrightarrow[\text{several stages}]{\text{fast}} CH_3\cdot CO_2H + 3EtOH \quad (78)$$

$$\textbf{48} \qquad\qquad\qquad \textbf{49}$$

$$\frac{-d[\textbf{48}]}{dt} = k_u[\textbf{48}] + k_h[H_3O^+][\textbf{48}] + k_n[HO\cdot C_6H_4\cdot NO_2][\textbf{48}]. \quad (79)$$

This kinetic form, with terms involving $[H_3O^+]$ and the m-nitrophenol concentration, suggests that the rate-determining step involves transfer of a proton from an acid molecule to the ethyl ortho-acetate (reaction 78a). This proton can most readily be supplied by H_3O^+ ($k_h \gg k_n$), but if the solution is not too acid then $[HO\cdot C_6H_4\cdot NO_2] \gg [H_3O^+]$ and a significant contribution to the rate from the k_n term can arise. The term k_u is seen by analogy probably to be due to water acting as an acid, i.e. the term in k_u in equation (79) should be replaced by $k'_u[H_2O]$ [48].

An example of a base-catalysed reaction is the self oxidation and reduction of aldehydes (the Cannizzaro reaction). In Chapter 2 evidence was presented that a hydrogen from one aldehyde CHO group is transferred directly to another aldehyde molecule [reaction (39) p. 19]. At low base concentration, the reaction is found to be of second order in aldehyde and first order in hydroxide ion concentration. If greater concentrations of hydroxide ion are used, the kinetic dependence becomes

$$-\frac{d[RCHO]}{dt} = k_1[RCHO]^2[OH^-] + k_2[RCHO]^2[OH^-]^2. \qquad (80)$$

These results are explained by scheme (81). At low concentrations of base, route (81a) is followed. Since an equilibrium is set up between RCHO and the anion **50**, a third-order rate expression

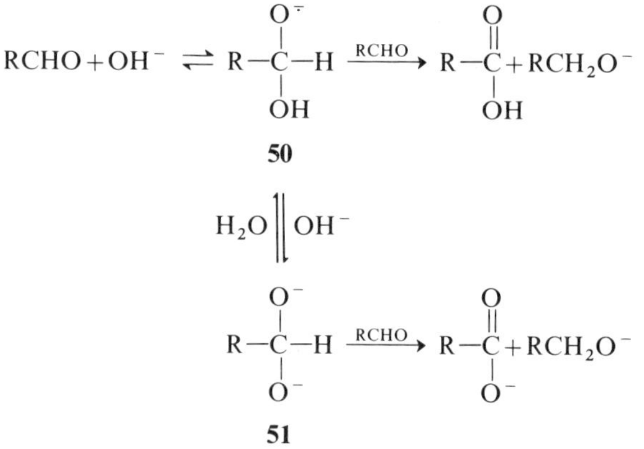

would be expected overall [compare reactions (66) and (67) in the previous section]. However, at higher base concentrations reaction of **50** with OH$^-$ can give **51**, opening up a new, overall fourth-order, route to the products (reaction 81b). Base- and acid-catalysed reactions will be considered further in Chapter 4, with particular emphasis on the intermediates like **49** and **50**.

Reactions of fractional order

Reactions which show truly fractional-order kinetics (i.e. not the mixed-order kinetics considered in the last section) often involve free-radical intermediates. Diagnostic tests for the presence of free-radical intermediates are described in the next chapter, but from the kinetic point of view, the most important feature of the majority of radicals is their great reactivity: if no suitable molecules are present with which they can react, they react with each other and are lost by combination and disproportionation reactions such as (82) and (83), which occur at rates approaching the collision frequency. As a consequence of the extreme rapidity of combination and disproportionation reactions, radical concentrations do not normally rise above about 10^{-8} M, and other reactions, if they are to compete with combination and disproportionation, must be fast.

$$2CH_3 \cdot CH_2 \cdot \begin{cases} CH_3 \cdot CH_2 \cdot CH_2 \cdot CH_3 & \text{combination} \quad (82) \\ CH_3 \cdot CH_3 + CH_2 : CH_2 & \text{disproportionation} \quad (83) \end{cases}$$

The steady-state approximation

Kinetic treatment of systems involving highly reactive intermediates such as free radicals is simplified by using this approach. If the concentration of an intermediate never becomes significant compared with the reagent and product concentrations, then after a short time the rates of production and destruction of the intermediate will become effectively equal and a 'steady state' is set up. The steady-state assumption does not imply that the concentration of the intermediate does not change during the reaction (in fact after infinite time in most reactions, the concentration of the intermediate will have died away to zero), but that at any particular time, the rates of production and destruction of the intermediate may be considered to be equal.

As a simple example of the use of the steady-state approximation, let us consider the decomposition of azoisobutyronitrile **52** in benzene solution. 2-Cyano-2-propyl radicals, **53**, are formed by reaction (84), and it is assumed for simplicity that they are lost only by reaction (85).

$$(CH_3)_2\underset{\substack{|\\CN}}{C}-N{=}N-\underset{\substack{|\\CN}}{C}(CH_3)_2 \;\rightarrow\; 2(CH_3)_2\underset{\substack{|\\CN}}{C}\cdot + N_2 \qquad (84)$$

$$\qquad \textbf{52} \qquad\qquad\qquad\qquad \textbf{53}$$

$$2(CH_3)_2\underset{\substack{|\\CN}}{C}\cdot \;\rightarrow\; (CH_3)_2\underset{\substack{|\\CN}}{C}-\underset{\substack{|\\CN}}{C}(CH_3)_2 \qquad (85)$$

Application of the steady-state approximation gives:

$$\frac{d[\mathbf{53}]}{dt} = 2k_{84}[\mathbf{52}] - 2k_{85}[\mathbf{53}]^2 = 0 \qquad (86)$$

or

$$[\mathbf{53}] = (k_{84}[\mathbf{52}]/k_{85})^{\frac{1}{2}} \qquad (87)$$

This is an important result: that the steady-state concentration of a radical is usually proportional to the square root of its rate of production.

It should be emphasized that the steady-state approximation can also be applied to reactions where reactive intermediates other than radicals are involved, for example, carbonium ions, carbanions, and carbenes.

Chain reactions

Most of the free-radical reactions commonly encountered are chain reactions. In a chain reaction, a radical produced in the system reacts with a molecule to give a new radical, which in turn is able to react with another molecule and so on. Finally the radical will be lost by combination or disproportionation, but before this happens a very large number of molecules (typically ~ 1000) will have reacted. The role of the original (or *initiating*) radical is similar to that of an acid or base molecule in an acid- or base-catalysed reaction. Typical free-radical chain reactions include substitutions and addition of molecules to unsaturated compounds.

A general scheme for a fairly simple chain reaction is given below. This scheme would fit, for example, the chlorination of alkanes ($X = $ Hal; $A = $ H; $B = $ R).

$$X-X \xrightarrow{\text{heat or } h\nu} 2X\cdot \qquad \text{initiation} \qquad (88)$$

$$X\cdot + A-B \longrightarrow X-A+B\cdot \left.\begin{array}{l}\\ \\\end{array}\right\} \begin{array}{l}\text{propagation; the overall effect} \\ \text{of these two reactions is}\end{array} \qquad (89)$$

$$B\cdot + X-X \longrightarrow B-X+X\cdot \quad X\cdot + AB + X_2 \ \rightarrow \ AX + BX + X\cdot \qquad (90)$$

$$X\cdot + X\cdot \longrightarrow X-X \left.\begin{array}{l}\\ \\\end{array}\right. \qquad (91)$$

$$X\cdot + B\cdot \longrightarrow X-B \ \Big\} \text{termination} \qquad (92)$$

$$B\cdot + B\cdot \longrightarrow B-B \left.\begin{array}{l}\\\end{array}\right. \qquad (93)$$

For a chain reaction in which (say) HBr was added to an olefin, a scheme similar in principle would be followed, but one of the propagation steps would involve addition of Br· to the olefin (94), followed by a hydrogen transfer reaction (95).

$$Br\cdot + CH_3-CH{=}CH_2 \ \rightarrow \ CH_3-\overset{\cdot}{C}H-CH_2Br \qquad (94)$$

$$CH_3-\overset{\cdot}{C}H-CH_2Br + HBr \ \rightarrow \ CH_3-CH_2-CH_2Br + Br\cdot \qquad (95)$$

Kinetic analysis of chain reactions is still complex, even if the steady-state approximation is used. However, in most chain reactions, it can be arranged that one of the propagation steps (89) or (90) is much faster than the other, either because of a difference in rate constants, or if these are similar, the presence of excess X–X or A–B produces the same effect. If reaction (89) is much faster than (90), the main radical present in the system is B·. Under these conditions, termination is mainly by (93) and the steady-state approximation gives expression (96). Conversely, if reaction (90) is much faster than (89), [X·] $\gg$ [B·], termination is mainly by (91), and the kinetics simplify to (97). These alternatives are readily distinguishable.

$$\frac{d[BX]}{dt} = \frac{k_{90}k_{88}^{\frac{1}{2}}}{k_{93}^{\frac{1}{2}}}[X_2]^{\frac{1}{2}} \qquad \boxed{[B·] \gg [X·]} \qquad (96)$$

$$\frac{d[BX]}{dt} = \frac{k_{89}k_{88}^{\frac{1}{2}}}{k_{91}^{\frac{1}{2}}}[X_2]^{\frac{1}{2}}[AB] \qquad \boxed{[X·] \gg [B·]} \qquad (97)$$

For a particular reaction, the kinetic order followed indicates which of the propagation reactions is the fast step, and from the composite rate constant, information about the rate-determining propagation step can be obtained.

For example, the reaction of bromine with chloroform to give hydrogen bromide and bromotrichloromethane in the gas phase at 150–180°C follows the kinetic expression:

$$-d[Br_2]/dt = k_{obs}[CHCl_3][Br_2]^{\frac{1}{2}} \qquad (98)$$

This suggests that the reaction is a radical chain reaction, with the two propagation steps (99) and (100).

$$Br· + CHCl_3 \longrightarrow HBr + ·CCl_3 \qquad (99)$$

$$·CCl_3 + Br_2 \longrightarrow BrCCl_3 + Br· \qquad (100)$$

Since expression (98) is of the type (97) rather than (96), reaction (100) is fast and (99) is the slower or rate-determining step, and bromine is effectively the only radical present.

By using known values for the equilibrium constant for bromine dissociation (k_{88}/k_{91}; X = Br) at a particular temperature, the rate constant for reaction (99) can be worked out from the overall rate. All that can be said about reaction (100) is that it must have a rate constant substantially greater than k_{99}. For reactions where k_{89} and k_{90} are of the same order of magnitude, both of the limiting rate-expressions (96) and (97) can be obtained by using a large excess of one or the other reagent, and information about both propagation steps can thus be obtained. Unless this can be done kinetic information will only be obtained about the slower of the two propagation reactions and the only deduction which can be made about the other one is that it is faster.

Reactions with even more complex dependence of rate on concentrations

Not all reactions can be expressed in terms of integral, fractional or mixed-order dependence on reagent concentrations. Two common reasons for more complex behaviour are:

(a) *The formation of intermediates in significant amounts*

If this happens, the rate of formation of final products will not be the same as the rate of disappearance of the reagents, and a complex kinetic dependence of rate of formation of products on reagent concentrations will be found. In cases of this sort it may be possible to measure the concentration of the intermediate, in which case the reaction kinetics may be analysable in terms of the two or more consecutive reactions taking place. Simple examples where behaviour of this general kind is found include the hydrolyses of dihalides and the esters of dicarboxylic acids.

(b) *One of the products inhibits the reaction*

In any reversible reaction where an appreciable concentration of reactants is left when equilibrium has been achieved, reverse reactions will become important after a short time when appreciable concentrations of products have been produced. Alternatively, the reaction may be inhibited (slowed down) by adding one or more of the products. As an example of inhibition, we may consider the bromination of chloroform [reactions (99) and (100)]. In this reaction, the reverse reaction (-99) has an appreciable rate constant, so that the presence of HBr in the system, either formed during the reaction or added at the outset, will reduce the rate of disappearance of bromine, and an expression of type (101) is obtained.

$$\frac{-\mathrm{d}[Br_2]}{\mathrm{d}t} = \frac{k_{obs}[CHCl_3][Br_2]^{\frac{1}{2}}}{1 + k_{-99}[HBr]/k_{100}[Br_2]} \tag{101}$$

This reduces to expression (98) for zero hydrogen bromide concentration.

Kinetic expressions of type (101) with a product concentration in the denominator provide an indication that inhibition by a product is taking place. Experimentally, it is usually best to simplify situations of this type by measuring initial rates (i.e. only covering the first few per cent of the reaction), in which case inhibition terms can be ignored, and a simpler expression such as (98) is obtained.

The dependence of reaction rate on temperature

Almost all reactions increase in rate with increasing temperature. For elementary reactions (and some complex reactions), it is found that the variation of reaction rate with temperature can normally be fitted within experimental error by an expression of type (102), known as an Arrhenius equation

$$k = A \exp(-E/RT) \tag{102}$$

In terms of the collision theory of reactions, the *activation energy*, E, is seen as the critical energy that a molecule needs in order to undergo a unimolecular reaction or, in a bimolecular reaction, the combined energy which two molecules need at the moment of collision for the reaction to take place. The Arrhenius *pre-exponential factor*, A for bimolecular reactions is seen as the rate at which effective collisions between molecules occur and is often dissected into two terms $A = PZ$, where Z is the collision frequency and P is the steric factor; to account for the fact that A is normally less than Z it may be assumed that all collisions are not effective, and that reaction can only take place if the molecules collide in particular orientations. The Arrhenius pre-exponential factor for unimolecular reactions is slightly more difficult to interpret. However, transition-state theory is now more commonly used to provide insight into both unimolecular and bimolecular reactions.

According to transition-state theory, reactants are thought of as being in a pseudo-equilibrium with a transition state which is the highest point on the potential energy diagram (Figure 2) followed when reactants go to products. The reaction coordinate in Figure 2 can be regarded as the distance (or a suitable arbitrary function of it) between any two atoms in the reacting system

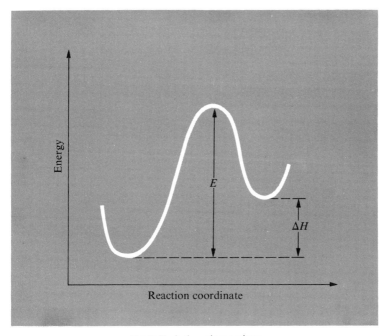

FIG 2. Endothermic reactions.

between which a covalent bond is being made or broken during the reaction. All other bond lengths and bond angles are to be regarded as being chosen to provide minimum energy at any point on the curve. Hence the transition state should be regarded as a col or saddle point, rather than as a peak in the potential-energy–reaction-coordinate diagram.

Using this approach, the following equation may be derived for the rate constant of a chemical reaction.

$$(kT/h) \exp(\Delta S^{\ddagger}/R) \exp(-\Delta H^{\ddagger}/RT) \qquad (103)$$

In this equation, k and h are Boltzmann's and Planck's constants respectively. $\Delta S^{\ddagger}$ and $\Delta H^{\ddagger}$ are the differences in entropy and enthalpy between the transition state and the reactants (e.g. $\Delta S^{\ddagger} = S_{\text{transition state}} - \Sigma S_{\text{reactants}}$). For some reactions it is possible to make *a priori* estimates of $\Delta S^{\ddagger}$ and $\Delta H^{\ddagger}$ which can then be compared with experimental kinetic results since for reactions in solution relationships (104) and (105) can be derived. (Slight modifications of these formulae are needed for gas-phase reactions).

$$\Delta H^{\ddagger} = E - RT, \qquad (104)$$

$$A = (kT/h) \exp(1 + \Delta S^{\ddagger}/R). \qquad (105)$$

Information from activation energies

The *a priori* estimation of activation energies is normally difficult. However, for endothermic reactions, the activation energy must be at least as great as the endothermicity (see Figure 2). In fact for very endothermic reactions (such as are often involved in the production of unstable intermediates) it is often found that the endothermicity and activation energy are approximately equal. Since endothermicities—which depend only on heats of formation of reactants and products—can often be calculated, this establishes minimum values for the activation energies of endothermic reactions. If the experimental value is found to be very much below this value, then the particular reaction mechanism can be ruled out of consideration.

As a very simple example of this approach, let us consider the thermal decomposition of tetraethyl lead, which gives lead and various organic products derived from the ethyl radicals produced in the decomposition.

$$PbEt_4 \xrightarrow{(106)} Pb + 4Et\cdot \longrightarrow \text{ethane, ethylene, butane.}$$

(107) ↓   ↗ one or more   (106, 107)

$$Et\cdot + PbEt_3\cdot \quad \text{stages}$$

Two possible routes to the observed products are shown above. Reaction (106) involves the simultaneous rupture of all four lead–carbon bonds. Reaction (107) involves breakage of only one lead–carbon bond initially, with subsequent breaking of the others. The subsequent reactions of the ethyl radicals to produce ethane, ethylene and butane are known to have virtually zero activation energies and will be very fast. Hence reaction (106), (107), or any other postulated primary step will be rate-determining.

From the known heats of formation of tetraethyl lead, lead, and the ethyl radical, reaction (106) can be estimated to be 515 kJ mol^{-1} endothermic. This endothermicity is much greater than the observed activation energy of 155 kJ mol^{-1}. Hence it must be concluded that reaction (106) is not taking place, and that some reaction such as (107) is the rate-determining step in the decomposition.

The use of activation energy arguments to rule out one of the two possible radical chain mechanisms for the chlorination of alkanes has been referred to in Chapter 1 (p. 9).

Rejection of unreasonable mechanisms in this way is usually done with hindsight—there is often good evidence from other sources that a different mechanism is operating. However, the author is prepared to stick his neck out on one example which has not been extensively studied. The Group IVB organometallic hydrides $R_3MH(M = Si, Ge, Sn, Pb)$ can be added to alkenes. For $M = Si$, Ge, or Sn under appropriate conditions the reaction is a radical chain process.

$$R_3M\cdot + CH_2{=}CHR' \rightarrow R_3M{-}CH_2{-}\overset{\cdot}{C}HR', \qquad (108)$$

$$R_3M{-}CH_2{-}\overset{\cdot}{C}HR' + HMR_3 \rightarrow R_3M{-}CH_2{-}CH_2R' + R_3M\cdot \qquad (109)$$

Organolead hydrides will also add to olefins, but in this case reaction (108) can be calculated, using bond dissociation energy values, to be ~ 88 kJ mol^{-1} endothermic. Assuming a reasonable value for the A-factor and no 'extra' activation energy, this would make the rate of reaction (108) far too small to compete with radical combination and disproportionation processes. Hence other mechanisms, for example a four-centre molecular mechanism (110), must be postulated for this reaction.

$$\begin{array}{c} R_3M{-\!\!-\!\!-\!\!-}H \\ \raisebox{1ex}{\vdots} \qquad \raisebox{1ex}{\vdots} \\ CH_2{=\!\!=}CHR' \end{array} \rightarrow R_3M{-}CH_2{-}CH_2R' \qquad (110)$$

Arrhenius A-factors

Arrhenius A-factors may be expected to be similar within a particular class of reactions since, on the collision picture, the steric requirements should not vary too much within a class, and in the transition-state theory the entropy difference between the reactants and the transition state should remain similar within a class. This accords with experiment: a very large number of elementary reactions have A-factors falling within the ranges shown in Table 3.

The differences between classes of reactions may be considered most simply from the entropy point of view. For unimolecular reactions, homolytic fission

TABLE 3

Arrhenius pre-exponential factors

Molecularity	Type of reaction	Example	$\log_{10} A$ $(s^{-1} \ or \ M^{-1} s^{-1})$
1	Homolysis	$C_2H_6 \rightarrow 2CH_3\cdot$	15–17
1	Four-centre process	$C_2H_5Br \rightarrow CH_2{=}CH_2$ $+ HBr$	12·5–14
2	Radical combination	$2CH_3\cdot \rightarrow C_2H_6$	9–10·5
2	Radical transfer or addition to a double bond (reagent an organic radical).	$CH_3\cdot + C_2H_6 \rightarrow CH_4$ $+ C_2H_5\cdot$ $CH_3\cdot + CH_2{=}CH_2$ $\rightarrow CH_3{-}CH_2{-}CH_2\cdot$	7–9
2	As above, but with a univalent atom as reagent.	$Cl\cdot + CH_4 \rightarrow CH_3\cdot + HCl$	10–11

into two (or more) fragments produces considerably greater freedom of movement (vibrational and rotational) in the transition state than in the ground-state molecule. Hence the entropy of activation is positive, and the A-factors are large. On the other hand, four-centre processes involve partial formation of two new bonds in the transition state giving a 'tighter' transition state with less entropy and hence lower A-factors than the unimolecular homolysis.

For bimolecular reactions, combinations of two radicals have high A-factors, corresponding to a transition state in which little bonding has developed between the two radicals. For addition and transfer reactions of radicals, the higher A-factors observed when univalent atoms are involved as reagents may be ascribed to the fact that atoms have no rotational entropy, and hence rotational entropy cannot be lost (by the atom) in going to the transition state.

Sufficient examples of other classes of reaction will probably be available in the near future to allow further additions to Table 3. It should be noted that no ionic examples are given. Ionic reactions often produce very large entropy changes by ordering solvent molecules (or allowing the order to collapse) during the reaction. This makes it more difficult to make generalizations about A-factors for ionic reactions, as opposed to free-radical and molecular reactions where interactions with a solvent (if present) are much weaker.

A-factor studies have already thrown light on some reactions. The pyrolysis of cyclobutane takes place by a first order process to give two molecules of ethylene. It was originally suggested that the reaction was a four-centre molecular reaction (111) since attempts to inhibit the reaction with nitric oxide or propene (which should react with radicals if formed) failed. However, the A-factor observed is $10^{15.6}$, which inspection of Table 3 shows to be more characteristic of a homolytic fission process than a true four-centre reaction.

Detailed calculations on entropies of activation reinforce this point and it is now thought that route (112) is more likely: the very rapid occurrence of the second step in which the biradical **54** breaks up into two ethylene molecules is thought to explain why propene and nitric oxide have no effect on the reaction.

$$
\begin{array}{ccccc}
\text{CH}_2-\text{CH}_2 & & \text{CH}_2=\text{CH}_2 & & \text{CH}_2=\text{CH}_2 \\
| \quad | & \rightarrow & \vdots \quad \vdots & \rightarrow & \\
\text{CH}_2-\text{CH}_2 & & \text{CH}_2=\text{CH}_2 & & \text{CH}_2=\text{CH}_2
\end{array}
\qquad (111)
$$

$$
\begin{array}{ccc}
\text{CH}_2-\text{CH}_2\cdot & \xrightarrow{\text{fast}} & \text{CH}_2=\text{CH}_2 \\
| & & \\
\text{CH}_2-\text{CH}_2\cdot & & \text{CH}_2=\text{CH}_2
\end{array}
\qquad (112)
$$

54

Distinction between possible mechanisms on A-factor grounds has not been very widely practised in the past. However, as more data on elementary reactions accumulate and it becomes possible to assign values or ranges to a greater number of types of chemical reaction, more use of this type of reasoning seems likely.

Primary kinetic isotope effect

A specialized use of kinetics in determining mechanism involves the use of isotopes, nearly always deuterium or tritium, at a position in the molecule where reaction occurs.

The attractive and repulsive forces in a molecule, being electrostatic in nature, will be unchanged if one hydrogen atom in the molecule is replaced by a deuterium atom, since the nuclear charges of the two isotopes are the same. However, the greater mass of the deuterium atom will give the deuterium compound less zero-point energy than the protium (hydrogen) compound. Thus the bond dissociation energy of the deuterium compound will be greater than that of the hydrogen compound by about 5 kJ mol^{-1}. In chemical reactions in which the C—D or C—H bond is at least partially broken, the activation energy for the deuterium compound would be expected to be greater (see Fig. 3), though any residual zero-point energy in the transition state will reduce the difference between E_D and E_H. If the rate-determining step in a reaction involves weakening of an X—H bond, then replacement of that particular hydrogen by deuterium will slow down the reaction. The ratio of rate constants k_H/k_D for the reaction of the normal molecule compared with that of the deuterated molecule is called the primary isotope effect for the reaction considered. The maximum value of the primary isotope effect will depend on the zero-point energy differences; for reactions involving C—H bonds k_H/k_D cannot be greater than about 7 at 25°C, and k_H/k_T cannot

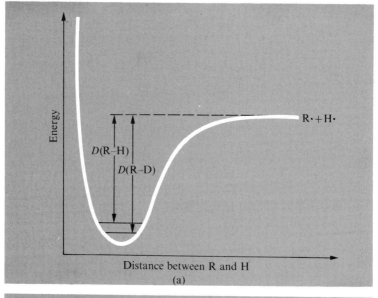

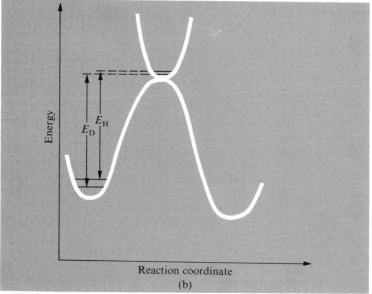

FIG. 3. The primary deuterium isotope effect. (a) The homolytic dissociation of an alkane showing the differences between the bond dissociation energies $D(R-H)$ and $D(R-D)$ due to zero-point energy differences (exaggerated). (b) Reaction in which some zero-point energy is retained in the transition state. $E_D - E_H$ is less than $D(R-D) - D(R-H)$.

be greater than about 17 unless other factors are involved. Somewhat higher values are possible for N—H and O—H bonds. A value for the primary isotope effect near the maximum indicates that the particular bond is almost completely broken in the transition state. Smaller values may indicate partial breakage of the bond in the transition state, while a value near unity implies that little or no breakage of the bond is taking place.

This technique has been used to support the view that in electrophilic aromatic substitution [e.g. the nitration reaction (113)] the mechanism involves addition of a nitronium ion to the aromatic nucleus to form the Wheland intermediate **55**, rather than the alternative synchronous process involving the transition state **56**.

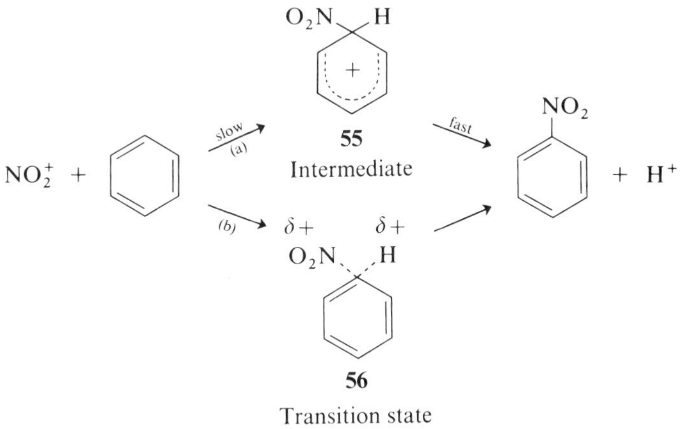

$$NO_2^+ + \quad\quad \xrightarrow[\;(a)\;]{slow} \quad\quad \textbf{55} \quad\quad \xrightarrow{fast} \quad\quad + H^+ \quad (113)$$

Intermediate

56

Transition state

In mechanism (a) there is no appreciable breaking of the C—H bond in the slow step to form the Wheland intermediate **55**, whereas in the alternative mechanism (b), the C—H bond is approximately half broken in the transition state **56**. Accordingly an appreciable isotope effect is expected for mechanism (b), but not for mechanism (a). Experimentally, benzene containing a very small proportion of the tritium isotope was converted into nitrobenzene and dinitrobenzene. Within experimental error, there was no enrichment of tritium in the nitrated products implying that reaction is as fast at C—T as at C—H positions, agreeing with mechanism (a) but not with mechanism (b).

An example of a reaction where an isotope effect *is* observed is the oxidation of alcohols by Cr(VI). In dilute acid solution, isopropanol is oxidized to acetone by the $HCrO_4^-$ ion at a rate proportional to $[alcohol][HCrO_4^-][H^+]$. The rate of oxidation is reduced by a factor of seven if the isopropanol is replaced by the deuteriated compound $(CH_3)_2CD \cdot OH$. Hence the carbon–deuterium bond in the deuteriated isopropanol must be partially broken in the transition state. This and other evidence supports the mechanism (114) in which an equilibrium is set up between the alcohol and the chromate

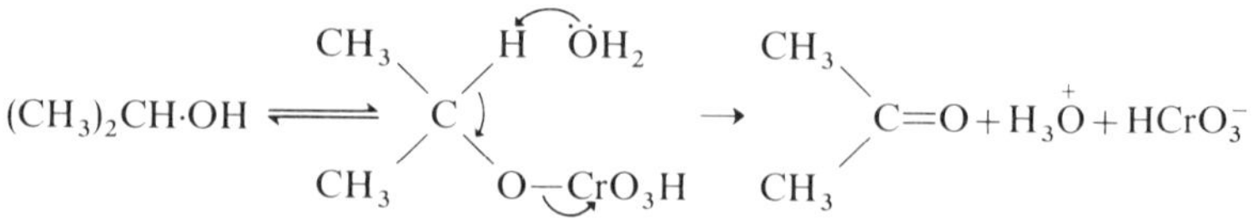

$$(CH_3)_2CH \cdot OH \rightleftharpoons \underset{57}{\overset{\displaystyle CH_3 \diagdown \quad \overset{\frown}{H} \quad \overset{\cdot \cdot}{O}H_2}{\underset{CH_3 \diagup \quad O - \underset{\smile}{C}rO_3H}{C}}} \longrightarrow \underset{CH_3 \diagup}{\overset{CH_3 \diagdown}{C}} = O + H_3\overset{+}{O} + HCrO_3^-$$

(114)

ester **57**, followed by a rate-determining removal of hydrogen (or deuterium) from **57** by a base (which is shown as water in the equation, but may be one of the chromate oxygen atoms under certain conditions), with concurrent formation of the carbonyl double bond and cleavage of the chromium–oxygen bond as shown. A similar isotope effect of about seven is found for the base-catalysed bromination of acetone, when the rate-determining stage is thought to be the removal of a proton by the base [reaction (115)]:

$$CH_3 \cdot CO \cdot CH_3 \xrightarrow[slow]{OH^-} CH_3 \cdot CO \cdot CH_2^- \xrightarrow[fast]{Br_2} CH_3 \cdot CO \cdot CH_2Br$$
$$(CD_3 \cdot CO \cdot CD_3)$$

(115)

Substitution of deuterium in the methyl groups of isopropyl bromide causes no appreciable change in the rate of its S_N2 reaction (116) with ethoxide ion, but the concurrent elimination [reaction (117)] is slowed by a factor of 6·7 on deuterium substitution.

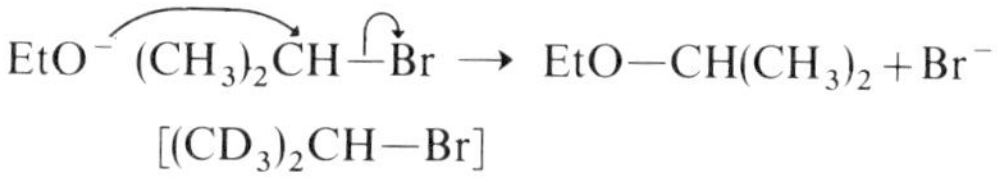

$$EtO^- \overset{\frown}{(CH_3)_2\overset{\frown}{CH}} \overset{\frown}{-} Br \rightarrow EtO - CH(CH_3)_2 + Br^-$$
$$[(CD_3)_2CH - Br]$$

(116)

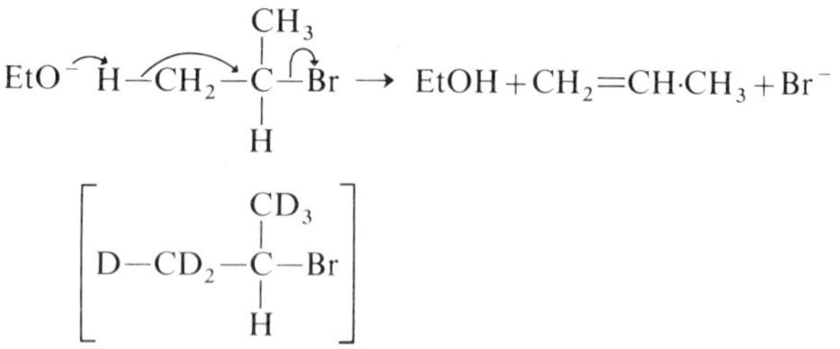

$$EtO^- \overset{\frown}{H} \overset{\frown}{-} CH_2 \overset{\displaystyle \overset{CH_3}{|}}{\underset{|}{C}} \overset{\frown}{-} Br \rightarrow EtOH + CH_2 = CH \cdot CH_3 + Br^-$$
$$\underset{H}{|}$$

(117)

$$\begin{bmatrix} & CD_3 & \\ & | & \\ D - CD_2 - & C - Br \\ & | & \\ & H & \end{bmatrix}$$

Use of isotopes of elements other than hydrogen

In principle, isotopes of elements other than hydrogen may be used in kinetic isotope effect studies. Among common elements found in organic compounds, the isotopes ^{13}C, ^{14}C, ^{15}N, and ^{18}O can be used. However, the maximum kinetic isotope effect which can be observed with any of these isotopes is only of the order of ten per cent or so (to be contrasted with the 700 per cent often observed for deuterium isotope effects), because replacement of (say) ^{12}C by ^{13}C in a compound produces a very much smaller percentage change in reduced mass and zero-point energy than does replacement of ^{1}H by ^{2}D. This means that extremely precise kinetic measurements are required to measure isotope effects for elements other than hydrogen, and so far relatively few examples of such isotope effects have been reported.

Summary

After the rate of a chemical reaction has been measured it is analysed in terms of its dependence on reactant concentrations. Conclusions about mechanism can be drawn depending on whether the reaction is of integral, fractional, or mixed order with respect to reactants, or shows more complex behaviour. When rate constants for elementary reactions can be deduced, reactions carried out at several temperatures give Arrhenius parameters, which themselves may give information about mechanism or allow a choice between mechanisms to be made. Finally, kinetic isotope effect studies can show whether a particular bond is broken in the rate-determining step.

PROBLEMS

3.1. How would you measure the rates of the following reactions?
   (a) $C_2H_5Br + OH^- \longrightarrow C_2H_5OH + Br^-$
   (b) $CH_3 \cdot CO_2 \cdot CHPh \cdot CH_3 \xrightarrow[\text{phase}]{\text{gas}} CH_3 \cdot CO_2H + PhCH{=}CH_2$
   (c) $CH_3 \cdot CO \cdot CH_3 + Br_2 \xrightarrow{H^+} CH_3 \cdot CO \cdot CH_2Br + HBr$

3.2. Write mechanisms which are consistent with the observations that:
   (a) solutions of dibenzylmercury in inert solvents decompose to give mercury and bibenzyl, showing first-order kinetics.
   (b) solutions of $(Me_3Si)_2Hg$ in inert solvents decompose to give $Hg + Me_3Si \cdot SiMe_3$, showing second-order kinetics.

3.3. The rate of nitration of benzene in a mixture of nitric acid and acetic acid does not depend on the benzene concentration. What can be deduced from this?

3.4. A simplified reaction scheme for the autoxidation of hydrocarbons by oxygen is shown below.

$$\text{Initiator} \longrightarrow 2R\cdot \qquad (1)$$

$$R\cdot + O_2 \xrightarrow{\text{fast}} RO_2\cdot \qquad (2)$$

$$RO_2\cdot + RH \xrightarrow{\text{slow}} R\cdot + RO_2H \qquad (3)$$

Termination is mainly by (4)

$$2RO_2\cdot \longrightarrow ROOR + O_2 \qquad (4)$$

Oxygen may be presumed to be present in excess. Apply the steady-state treatment to work out the expected kinetics for the reaction. Why is the termination step (4), which involves two $RO_2\cdot$ radicals, likely to be much more significant than alternative termination reactions?

3.5. Suggest a mechanism or mechanisms for the benzidine rearrangement (**A** $\longrightarrow$ **B**), which is consistent with the following information (1) the reaction is first order with respect to **A** and second order with respect to $[H^+]$, (2) the reaction is strictly intramolecular, and (3) there is no isotope effect if the *para* H atoms are replaced by D.

A **B**

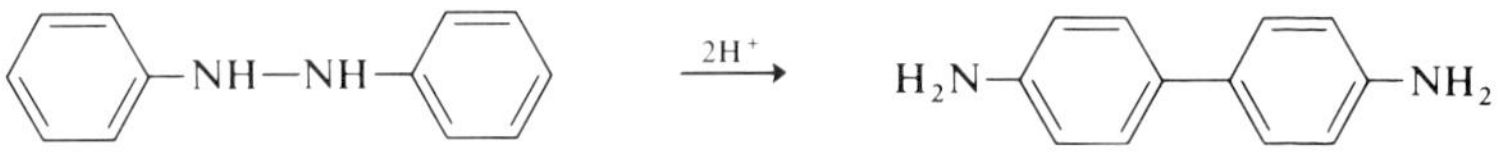

3.6. The Arrhenius *A*-factor for the first-order gas-phase pyrolysis of ethyl acetate is $10^{12.5}$. Which of the two postulated mechanisms does this piece of evidence favour?

(a)

$$CH_3{\cdot}CO_2{\cdot}C_2H_5 \xrightarrow{\text{slow}} CH_3{\cdot}CO_2\cdot + \cdot C_2H_5 \xrightarrow{\text{fast}} CH_3{\cdot}CO_2H + C_2H_4 \qquad \text{(b)}$$

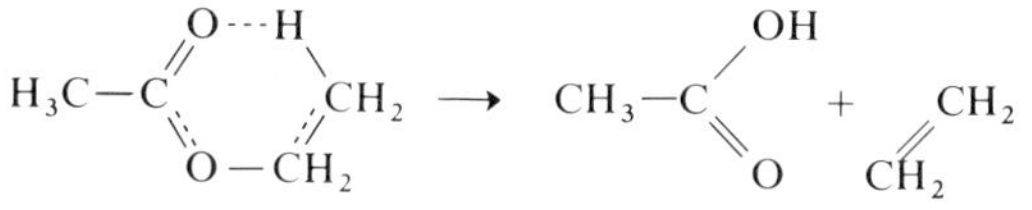

4. Intermediates

A MINORITY of observed chemical reactions are elementary: the reactant molecules (not more than two, or three at the most) react in one step to give product molecules. More commonly, the reaction is complex and involves two or more elementary reactions, in which case one or more intermediate species must have been formed. Evidence from kinetics considered in the last chapter often indicates or proves that intermediates of some kind have been produced in a reaction and may give an indication of the type of intermediate involved.

In this chapter we shall consider techniques for establishing in more detail the precise nature of the intermediates formed in chemical reactions. These techniques are of two kinds: *Spectroscopic techniques*, particularly n.m.r., e.s.r., u.v., and mass spectroscopy may be used to provide evidence for intermediates; alternatively the intermediates may be *trapped*, occasionally by the isolation of the intermediate itself, but more usually by adding to the reacting system a compound which will react with the postulated intermediate, and looking for the product of such a reaction in the final work up. Various types of intermediates are encountered in organic reactions: these include carbonium ions, carbanions, radicals, carbenes, benzynes, tetrahedral intermediates, and stable molecules (not listed in order of importance). It is convenient to consider them separately.

Carbonium ions, radicals (carbon centred), and carbanions

These three types of intermediate all contain three groups attached to a central carbon atom. The fourth, non-bonding orbital contains 0, 1, or 2 electrons for the three types of intermediate respectively. The polar consequences of the differing number of non-bonding electrons will be considered in more detail in Chapter 6, and the stereochemical consequences of formation of these intermediates in Chapter 5. Here it is sufficient to mention that the formation of any of these intermediates at an asymmetric centre may lead to loss of optical activity; hence loss of optical activity during a reaction is an indication that one of these types of intermediates may have been involved, but does not indicate which type.

All three types of intermediate will initiate the polymerization of suitable olefins. If an olefin is added to a reaction mixture in which (say) radicals are being formed, a chain reaction takes place in which successive olefin molecules add on to the initiating radical, giving rise eventually to a polymer molecule which comprises a large number of the original olefin units.

$$R\cdot + CH_2{=}CHR' \;\rightarrow\; R{-}CH_2{-}\dot{C}HR' \xrightarrow{\;CH^2{=}CHR'\;}$$

$$\text{(118)}$$

$$R{-}CH_2{-}CHR'{-}CH_2{-}\dot{C}HR' \;\rightarrow\; \rightarrow\; \text{polymer}$$

The polymer either precipitates during the reaction, or can be precipitated by methanol after the reaction is complete, and is therefore easily identified. Not all olefins (monomers) are polymerized with equal facility by cations, radicals, and anions. For example, methyl methacrylate is polymerized by anions and radicals, but not by cations, whereas styrene is polymerized by radicals and cations, but not by anions. The preferential attack by anions on methyl methacrylate rather than on styrene is probably a direct result of the negative charge on the anion. If attack by an anion takes place on the CH_2 group of methyl methacrylate, the resultant anion **58** will be stabilized by delocalization of the negative charge onto the oxygen atom **59**; this stabilization will be retained if the anion **58** attacks another molecule of methyl methacrylate, but would be lost if say a styrene molecule were attacked.

$$R^- + CH_2{=}\underset{\underset{CH_3}{|}}{C}{-}\overset{\overset{O}{\|}}{C}{-}OCH_3 \;\rightarrow\; R{-}CH_2{-}\underset{\underset{CH_3}{|}}{\overset{-}{C}}{-}\overset{\overset{O}{\|}}{C}{-}OCH_3 \;\leftrightarrow\; R{-}CH_2{-}\underset{\underset{CH_3}{|}}{C}{=}\overset{\overset{O^-}{|}}{C}{-}OCH_3$$

$$\textbf{58} \qquad\qquad \textbf{59} \qquad \text{(119)}$$

The differing reactivity of different olefins towards the different types of initiators has been developed into a technique for determining which type of reactive intermediate is involved. An equimolar mixture of styrene and methyl methacrylate is added to the reaction system and when the reaction is over the polymer is precipitated with methanol and analysed by combustion to determine its composition. Anions react to give a polymer consisting exclusively of methyl methacrylate units; cations give a polymer comprising styrene units only. Radicals give a co-polymer which consists of approximately equal numbers of styrene and methyl methacrylate units. Table 4 shows

TABLE 4

Co-polymerization of methyl methacrylate and styrene in the presence of various initiators

Polymer composition	% Carbon	% Hydrogen	Initiator
Polystyrene	92·3	7·7	Cation
Poly(methyl methacrylate)	71·4	9·6	Anion
1:1 co-polymer of styrene and methyl methacrylate	82·9	8·6	Radical

that the composition of the polymer, and hence the nature of the initiator, can readily be determined from the combustion analysis results, since analytical figures are generally obtained to within ± 0.2 per cent quite easily.

An example of the use of this technique is in the investigation of the nature of the reactive species formed when olefins are irradiated with γ-rays. It is found that provided the olefins used are extremely dry, cationic polymerization takes place, but that if even a trace of water is present, radicals are responsible for the initiation. This was taken to mean that both cations and radicals (and anions) are being formed, but that the cations are very much more reactive in initiating polymerization. In the presence of traces of water the cations are destroyed but the radicals are unaffected, so that the mechanism changes to a slower radical polymerization.

We now turn to techniques and evidence for particular types of intermediate.

Carbonium ions

Carbonium ions are involved as intermediates in many important organic reactions. Their production in S_N1 reactions by direct ionization of halides in polar solvents has already been referred to. Metal salts such as mercuric chloride and ferric chloride assist such ionization by complexing with the halide ion. Other reactions in which carbonium ions are generated include electrophilic additions to olefins, the dehydration of alcohols by acids, and the reaction of amines with nitrous acid.

N.m.r. spectroscopy

When aliphatic fluorides are dissolved in antimony pentafluoride, a reversible reaction leads to the formation of carbonium ions in high enough concentration to allow their n.m.r. spectra to be recorded.

$$RF + SbF_5 \rightleftarrows R^+ + SbF_6^- \tag{120}$$

The spectra can be unequivocally assigned to the carbonium ion: the absorptions are all downfield compared with aliphatic halides, as would be expected for the less shielded carbonium ions. The t-butyl cation shows a single peak at $\delta = 4.35$ p.p.m., whereas the isopropyl cation shows a doublet at $\delta = 5.03$ p.p.m. and a septet at $\delta = 13.5$ p.p.m., corresponding to the methyl and C—H protons respectively. These n.m.r. experiments also confirm

(a) that stability of carbonium ions rises in the order primary < secondary < tertiary, and

(b) that carbonium ion rearrangements readily take place, since for example all the butyl fluorides (1-butyl, 2-butyl, i-butyl, and t-butyl) dissolve in antimony pentafluoride to give exclusively the t-butyl cation.

Trapping

Carbonium ions are readily trapped by nucleophiles, which may include solvents such as ethanol, or added nucleophiles such as halide or azide ions. The addition of bromine to ethylene (in the absence of radical initiators) is thought to proceed by the mechanism shown in (121). Evidence for the intermediacy of the carbonium ion **60** is provided by trapping experiments in which chloride and nitrate ions were added to the reaction mixture, resulting in the formation of **61** and **62** along with the normal product, 1,2-dibromoethane.

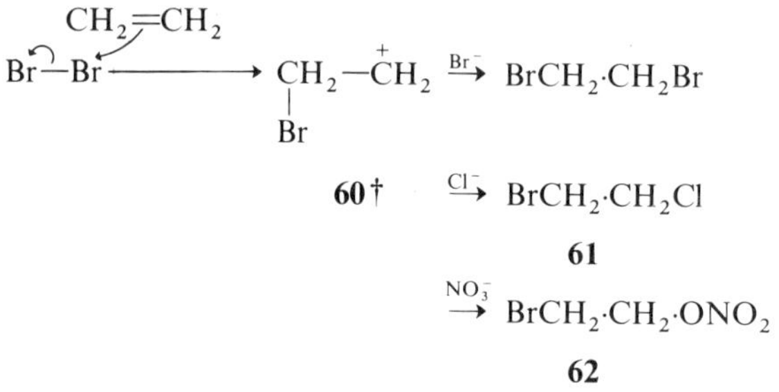

$$CH_2{=}CH_2$$
$$Br{-}Br \longrightarrow CH_2{-}\overset{+}{C}H_2 \overset{Br^-}{\to} BrCH_2{\cdot}CH_2Br$$
$$\underset{Br}{|}$$

$$\mathbf{60}\dagger \quad \overset{Cl^-}{\to} BrCH_2{\cdot}CH_2Cl \qquad (121)$$
$$\mathbf{61}$$
$$\overset{NO_3^-}{\to} BrCH_2{\cdot}CH_2{\cdot}ONO_2$$
$$\mathbf{62}$$

Solvolysis of the diester **63** has been postulated to involve the intermediate carbonium ion **64**. Evidence for the intermediacy of **64** is provided by the fact that it can be trapped by ethanol to give the ortho-ester **65**.

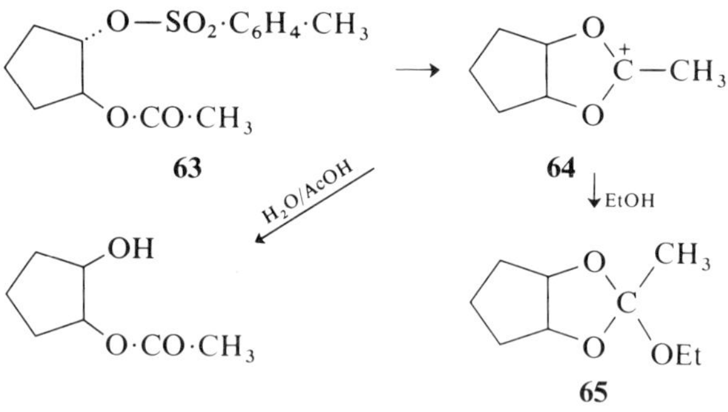

$$(122)$$

The diversion of the alcohol product during the solvolysis of alkyl halides by the addition of other nucleophiles without affecting the rate of the solvolysis was considered in the last chapter, reaction (59). Another logical consequence of carbonium-ion formation is that if two final products are given by a single carbonium-ion intermediate, the ratio of concentrations of the two products should be unchanged however the carbonium ion is produced. Thus although

† See p. 94 for a further discussion of the structure of this cation.

the decomposition of benzhydryl bromide **66** in 90 per cent aqueous acetone at 50°C in the presence of 0·101 M sodium azide is 335 times faster than the decomposition of the corresponding chloride **67**, the proportion of alcohol **68** to azide **69** is the same, within experimental error, pointing to a common intermediate—the carbonium ion **70**—in both solvolyses.

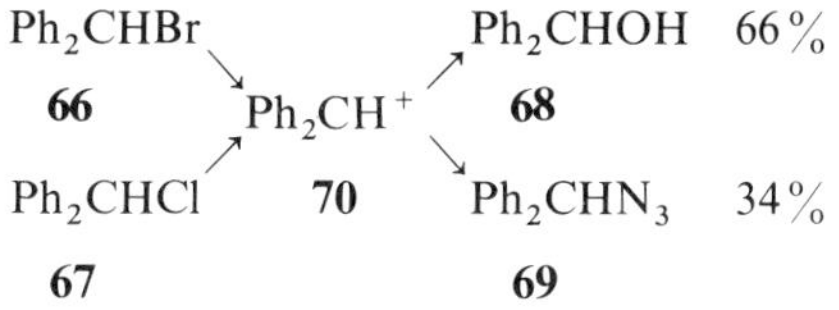

$$Ph_2CHBr \quad \quad Ph_2CHOH \quad 66\%$$
$$\mathbf{66} \quad \searrow Ph_2CH^+ \nearrow \quad \mathbf{68}$$
$$\nearrow \quad \quad \searrow \quad \quad \quad \quad \quad (123)$$
$$Ph_2CHCl \quad \mathbf{70} \quad Ph_2CHN_3 \quad 34\%$$
$$\mathbf{67} \quad \quad \quad \mathbf{69}$$

Rearrangements

An important characteristic of carbonium ions is that they readily undergo skeletal rearrangements, in which there is migration of an aryl or alkyl group or a hydrogen atom attached to the carbon atom next to the carbonium-ion site. This is shown in (124).

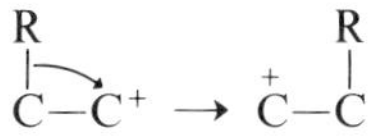

$$\overset{R}{\underset{|}{C}}-C^+ \rightarrow \overset{+}{C}-\overset{R}{\underset{|}{C}} \quad \quad (124)$$

Well-known reactions which involve a carbonium-ion rearrangement include the Wagner–Meerwein rearrangement, exemplified by (125), and the pinacol–pinacolone rearrangement (126).

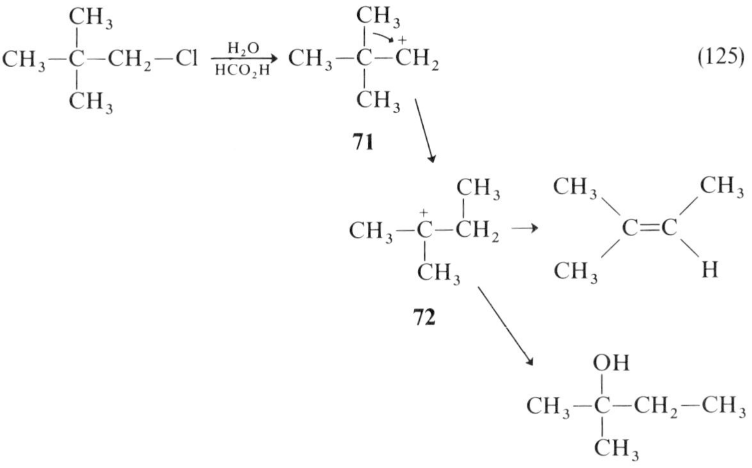

$$CH_3-\underset{\underset{CH_3}{|}}{\overset{\overset{CH_3}{|}}{C}}-CH_2-Cl \xrightarrow[HCO_2H]{H_2O} CH_3-\underset{\underset{CH_3}{|}}{\overset{\overset{CH_3}{|}}{C}}{}^+-CH_2 \quad (125)$$

71

$$CH_3-\overset{+}{\underset{\underset{CH_3}{|}}{\overset{\overset{CH_3}{|}}{C}}}-CH_2 \rightarrow \overset{CH_3}{\underset{CH_3}{}} C=C \overset{CH_3}{\underset{H}{}}$$

72

$$CH_3-\underset{\underset{CH_3}{|}}{\overset{\overset{OH}{|}}{C}}-CH_2-CH_3$$

$$CH_3-\overset{\overset{\displaystyle CH_3}{|}}{\underset{\underset{\displaystyle OH}{|}}{C}}-\overset{\overset{\displaystyle CH_3}{|}}{\underset{\underset{\displaystyle OH}{|}}{C}}-CH_3 \xrightarrow{H^+} CH_3-\overset{\overset{\displaystyle CH_3}{|}}{\underset{\underset{\displaystyle {}^+OH_2}{|}}{C}}-\overset{\overset{\displaystyle CH_3}{|}}{\underset{\underset{\displaystyle OH}{|}}{C}}-CH_3 \rightarrow CH_3-\overset{\overset{\displaystyle CH_3}{|}}{\underset{+}{C}}-\overset{\overset{\displaystyle CH_3}{|}}{\underset{\underset{\displaystyle OH}{|}}{C}}-CH_3$$

73 (126)

$$CH_3-\overset{\overset{\displaystyle CH_3}{|}}{\underset{\underset{\displaystyle CH_3}{|}}{C}}-CO-CH_3 \xleftarrow{-H^+} CH_3-\overset{\overset{\displaystyle CH_3}{|}}{\underset{\underset{\displaystyle CH_3}{|}}{C}}-\overset{+}{\underset{\underset{\displaystyle OH}{|}}{C}}-CH_3 \leftrightarrow CH_3-\overset{\overset{\displaystyle CH_3}{|}}{\underset{\underset{\displaystyle CH_3}{|}}{C}}-\underset{\underset{+}{\underset{\displaystyle OH}{|}}}{\overset{\|}{C}}-CH_3$$

74a   **74b**

In reaction (125), the primary neopentyl carbonium ion **71** rearranges to the more stable tertiary ion **72**. In reaction (126), the first-formed carbonium ion **73** is tertiary, but the rearranged cation **74** will have additional stabilization because of the charge delocalization. The isomerization of n-propyl to isopropyl cation which is observed when n-propyl fluoride is dissolved in antimony pentafluoride (see previous section) must be attributed to hydride migration (127).

$$CH_3-\overset{\overset{\displaystyle H}{|}}{\underset{\underset{\displaystyle H}{|}}{C}}\overset{+}{-}CH_2 \rightarrow CH_3-\overset{+}{\underset{\underset{\displaystyle H}{|}}{C}}-CH_3 \qquad (127)$$

Aryl groups migrate even more readily than alkyl groups or hydride.

Rearrangements are most uncommon when carbanion or radical intermediates are involved in a reaction. Hence if skeletal rearrangements have taken place during a reaction, this is evidence (though not conclusive evidence) that a carbonium-ion intermediate has been involved. Conversely, since carbonium-ion rearrangements are so facile, if a reaction is postulated to go via a carbonium ion which has an opportunity for rearrangement (i.e. neighbouring groups which can migrate to give a cation of equal or greater stability), and rearrangement products are not observed, then free carbonium ions have probably not been formed.

Carbanions

Carbanions are most commonly encountered as intermediates in reactions where a strong base can abstract a proton from a C—H bond in an organic molecule. Hydrocarbons are inert to such attack, but groups such as $\diagdown\hspace{-0.3em}C\hspace{-0.3em}O\diagup$ and $-CO_2Et$ activate adjacent C—H bonds and make attack by bases easier. The activating effect of the ketone or ester groups is seen as the result of stabilization of the resulting carbanion by delocalization of the charge (e.g. **75**). Two ketone or ester groups are more effective than one: although the equilibrium (128) lies well over to the left, the equilibrium (129) lies much further over to the right, and in fact **76** can be obtained in the form of its sodium salt.

$$CH_3 \cdot CO_2 \cdot Et + EtO^- \rightleftharpoons EtOH + \overset{\displaystyle O}{\overset{\displaystyle \|}{\bar{C}H_2 - C} - OEt} \longleftrightarrow CH_2 {=} \overset{\displaystyle O^-}{\overset{\displaystyle |}{C} - OEt} \quad (128)$$

$$ \mathbf{75a} \mathbf{75b}$$

$$CH_3 \cdot CO \cdot CH_2 \cdot CO_2 \cdot Et + EtO^-$$

$$\rightleftharpoons EtOH + CH_3 - \overset{\displaystyle O}{\overset{\displaystyle \|}{C}} - \overset{\displaystyle O}{\overset{\displaystyle \|}{\bar{C}H - C}} - OEt \longleftrightarrow CH_3 - \overset{\displaystyle O}{\overset{\displaystyle \|}{C}} - CH {=} \overset{\displaystyle O^-}{\overset{\displaystyle |}{C}} - OEt$$

$$\updownarrow \quad \mathbf{76} \hspace{7cm} (129)$$

$$CH_3 - \overset{\displaystyle O^-}{\overset{\displaystyle |}{C}} {=} CH - \overset{\displaystyle O}{\overset{\displaystyle \|}{C}} - OEt$$

A vinyl group also appears to stabilize a carbanion centre, though not to the same extent as a keto group. When either 1,3-cyclohexadiene or 1,4-cyclo-hexadiene is heated at 95°C in a mixture of t-C_5H_{11}—OK and t-C_5H_{11}—OD for a time sufficient for only a small amount of reaction to take place [reaction (130)], a mixture of monodeuteriated compounds **77** and **78** is formed in the same ratio of 4:1. Since this ratio is not the equilibrium mixture, this points to a common intermediate formed from the two starting materials, and this is almost certainly the carbanion **79**.

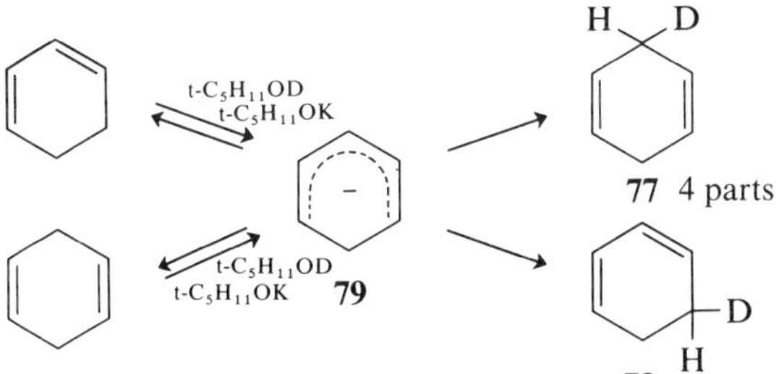

Carbanions where the negative charge is on an sp-hybridized carbon atom, rather than on an sp^2 or sp^3 carbon, are also stabilized, since sp-hybridized carbon is more electronegative than sp^2- or sp^3-hybridized carbon, and can therefore better accommodate a negative charge. Thus acetylenes with a terminal —C≡CH group are easily converted into their sodium salts. Carbanions are also found in reactions of organometallic compounds.

The main way in which carbanions are trapped is by addition to polarized multiple bonds. Thus carbanions initiate the polymerization of methyl methacrylate more readily than that of styrene (see the general section at the beginning of this chapter, page 53). Carbanions also readily add to the positively-polarized carbon atoms of carbonyl groups. Thus the aldol condensation is believed to involve the addition of a carbanion to an aldehyde or ketone molecule (131).

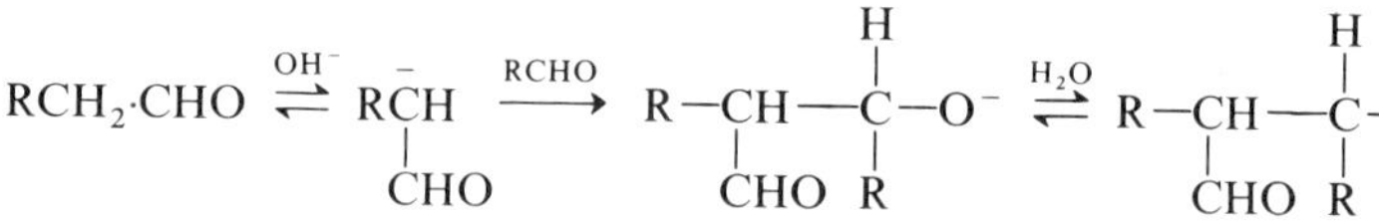

$$RCH_2 \cdot CHO \underset{}{\overset{OH^-}{\rightleftharpoons}} R\overset{-}{C}H \underset{|}{\overset{RCHO}{\longrightarrow}} \underset{CHO\ R}{R-CH-\overset{H}{\underset{|}{C}}-O^-} \overset{H_2O}{\rightleftharpoons} \underset{CHO\ R}{R-CH-\overset{H}{\underset{|}{C}}-OH}\ (131)$$

Carbanions are not obviously involved in the decarboxylation of pyridine and quinoline α-carboxylic acids such as picolinic acid **80**. However, the ready decarboxylation of these acids—relative to the β- and γ-acids—suggests that the nitrogen atom of the heterocyclic nucleus may remove a proton from the neighbouring carboxylic acid group to give the zwitterion **81**. This could plausibly lose carbon dioxide to give the carbanion **82** which would rapidly tautomerize to pyridine **83** by a proton shift. The intermediacy of the carbanion **82** was neatly shown by carrying out the reaction in the presence of acetophenone, when the adduct **84** was formed by addition of the carbanion to the carbon atom of the carbonyl group.

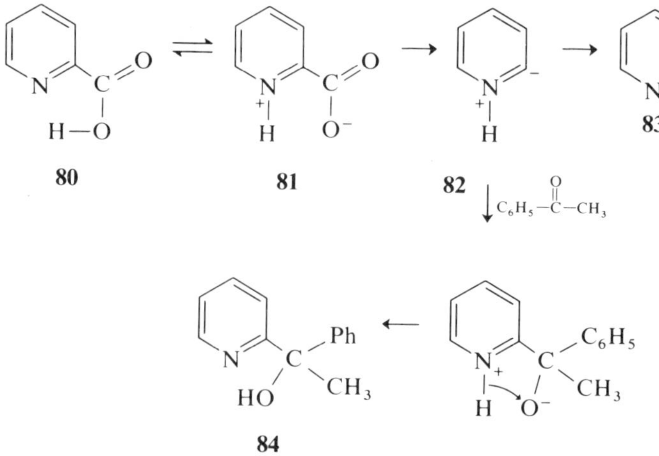

(132)

84

Radicals

Free-radical intermediates are often encountered in reactions carried out at high temperatures, particularly in the gas phase or in non-polar solvents. Radicals may be introduced into reaction systems by photolysis or pyrolysis of a suitable molecule, as discussed in the last chapter. Kinetic evidence for free-radical intermediates often takes the form of fractional orders of reaction with respect to reagent concentrations.

Spectroscopy

Because of their extremely rapid loss by combination reactions, radical intermediates are seldom encountered in concentrations above about 10^{-8} M. This rules out ultraviolet, infrared, and n.m.r. spectroscopy in most cases for detecting free-radical intermediates. However, *electron spin resonance* spectroscopy, which is very sensitive and will detect radical concentrations down to about 10^{-9} M, is extremely useful, since molecules without an unpaired electron are invisible to this form of spectroscopy. An electron spin resonance signal therefore is conclusive evidence that a free radical is present in the system, though of course it does not prove that the radical lies on a particular reaction path. For simple organic radicals a unique structural assignment can often be made from the hyperfine splitting pattern of the radical, or the spectrum can be compared with the known e.s.r. spectrum of the radical if this has already been obtained in another way.

The autoxidation of hydrocarbons by atmospheric oxygen is now believed to be a radical chain reaction which gives the hydroperoxide as the first stable product. The reaction is illustrated for isopropylbenzene (cumene) **85**. Radicals are probably produced in the system by reaction (133), and a chain reaction (134 and 135) is responsible for converting the hydrocarbon into its hydroperoxide.

$$PhCMe_2H + O{=}O \;\rightarrow\; PhCMe_2{\cdot} + {\cdot}OOH \qquad (133)$$

$$\mathbf{85} \qquad\qquad\qquad \mathbf{86}$$

$$PhCMe_2{\cdot} + O{=}O \;\rightarrow\; PhCMe_2{-}O{-}O{\cdot} \qquad (134)$$

$$\mathbf{87}$$

$$PhCMe_2{-}O{-}O{\cdot} + PhCMe_2H \;\rightarrow\; PhCMe_2{-}O{-}OH + PhCMe_2{\cdot} \qquad (135)$$

$$\mathbf{88}$$

When this reaction was carried out in the cavity of an e.s.r. spectrometer, a single-line signal attributable to the radical **87** developed. Radical **86** would give a very complex spectrum, so that the observed result both supports a radical chain mechanism for the autoxidation and implies that reaction (134) is rapid compared with (135), since otherwise the spectrum of **86** would be seen as well as or instead of that of **87**.

CIDNP (chemically induced dynamic nuclear polarization)

Free radicals cannot normally be detected by n.m.r. because of their low concentration and their paramagnetism, which leads to very broad lines. However, products derived from radical reactions may obviously be expected to show characteristic n.m.r. spectra. Since the unpaired electron, by its magnetic moment, can polarize the nuclear spins of the protons in the radical, the *products* of the radical reaction may have a distribution among the nuclear energy levels which differs markedly from the normal Boltzmann distribution. This gives rise to significant differences in intensity of the observed n.m.r. spectrum. If the lower energy levels are over-populated, enhanced absorption of radiofrequency energy takes place, while if the upper levels are over-populated, emission of energy occurs, and negative peaks are seen in the n.m.r.

CIDNP is still at the time of writing in a development stage: most of the reactions which have been studied are already known to involve free-radical intermediates. However, one example where CIDNP has suggested a radical mechanism where an ionic mechanism had previously been accepted is the thermal rearrangement of the compound **89** to the more stable non-charged structure **90**. Before CIDNP studies, the rearrangement was thought to take place either in a concerted manner, or via a 'tight' ion pair (reaction 136). However, CIDNP studies show a strong enhancement effect on the benzyl protons of the product **90**, suggesting that the rearrangement involves the homolysis of the nitrogen–carbon bond, followed by a cage recombination of the radicals as shown (reaction 137).

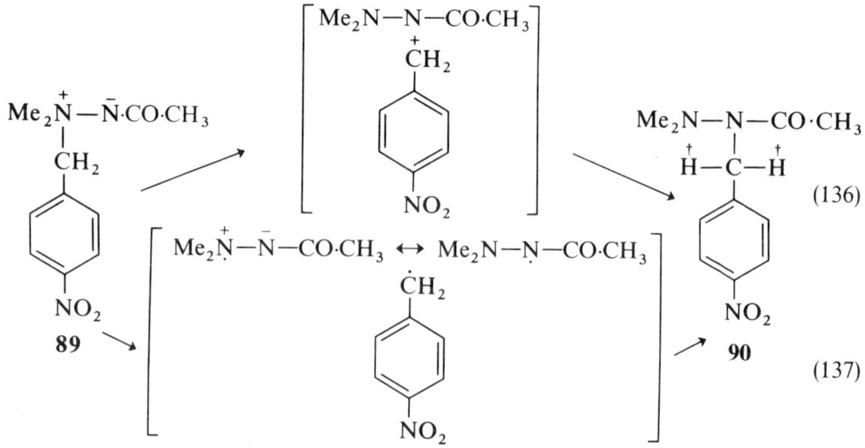

† CIDNP enhancement of these protons observed.

CIDNP shows considerable promise as a method for detecting and studying radical intermediates.

Trapping and related techniques

The presence of free radicals in a system is often indicated by the presence of small quantities of their dimers among the products. The occurrence of C_2F_6 as a by-product of the fluorination of methane has already been cited as evidence for the intermediacy of $CF_3\cdot$ radicals in the reaction. Conversely, the absence of a dimer is an indication that the corresponding radical may not have been present (though it must be remembered that in chain reactions the amounts of combination products formed may be extremely small). For example, phenyl radicals, produced by heating benzoyl peroxide, attack aromatic compounds to give the phenylated products **91**. The absence of the bi-aryl compound Ar–Ar among the products is evidence against route (138) which would involve Ar· radicals, and favours the alternative reaction scheme (139). Further evidence for scheme (139) is provided by the fact that dimers of the intermediate radical **92** can sometimes be isolated.

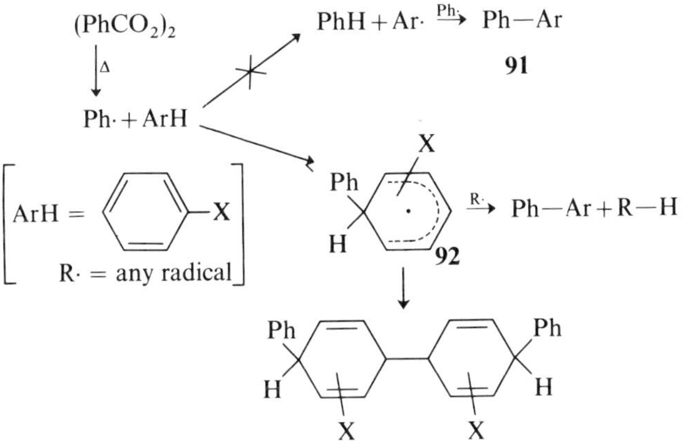

$$(PhCO_2)_2$$

$$PhH + Ar\cdot \xrightarrow{Ph\cdot} Ph-Ar \qquad (138)$$

$$91$$

$$Ph\cdot + ArH$$

$$ArH = \text{(benzene ring)}-X$$

$$R\cdot = \text{any radical}$$

$$\xrightarrow{R\cdot} Ph-Ar + R-H \qquad (139)$$

92

Free radicals may be trapped by the addition to the reaction system of a compound which readily reacts with free radicals, for example, bromine or nitric oxide. Thus when azoisobutyronitrile (**93**, R = $(CH_3)_2C(CN)-$) is heated in solution in the presence of nitric oxide, the hydroxylamine derivative **94** is formed, presumably by addition of three 2-cyano-2-propyl radicals to a nitric oxide molecule.

$$R-N{=}N-R \xrightarrow{\Delta} R\cdot + N_2 + R\cdot \qquad NO + 3R\cdot \rightarrow \rightarrow \ \substack{R \\ \diagdown \\ N-O-R} \qquad (140)$$

$$\substack{/ \\ R}$$

93 **94**

However, it is not always easy to isolate such adducts from a reaction mixture. One method of avoiding the need for product isolation is the technique of spin trapping. This involves addition to the reacting system of a molecule which will react with a reactive free radical to give a less reactive radical. Such a molecule is 2-methyl-2-nitroso-propane **95**, which adds to unstable radical intermediates R· to give the relatively stable nitroxide radical **96**, whose e.s.r. spectrum can be measured. From the hyperfine structure observed, information about the structure of R· can be deduced.

$$Bu^t-N{=}O + R\cdot \longrightarrow \begin{array}{c} Bu^t \\ \diagdown \\ N-O\cdot \\ \diagup \\ R \end{array} \tag{141}$$

$$\mathbf{95} \qquad\qquad \mathbf{96}$$

The important technique of detecting the presence of radicals in a system by adding a polymerizable olefin and looking for polymer has been referred to earlier in the chapter (p. 53).

Inhibition

In radical chain reactions, for example reactions (99) and (100) on page 41, the propagation steps involve conversion of reactant molecules to product molecules, whilst the initiating radical is re-generated. If a compound is present which reacts with the initiating radical to give either a neutral molecule [e.g. reaction (142)], or a radical which is too stabilized to carry on the reaction chain [e.g. **97** in reaction (143)], then the rate of the chain reaction will be reduced, or it may be stopped altogether. Compounds, termed inhibitors, which act in this way, include nitric oxide, olefins, phenols, aromatic amines, and nitro-compounds.

$$R\cdot + N{=}O \longrightarrow R-N{=}O \tag{142}$$

$$R\cdot + \underset{\underset{\displaystyle OH}{|}}{\overset{\overset{\displaystyle OH}{|}}{\bigcirc}} \longrightarrow R-H + \underset{\underset{\displaystyle OH}{|}}{\overset{\overset{\displaystyle O\cdot}{|}}{\bigcirc}} \longleftrightarrow \underset{\underset{\displaystyle OH}{|}}{\overset{\overset{\displaystyle O}{\|}}{\bigcirc}}{}^{\bullet} \text{ etc.} \tag{143}$$

$$\mathbf{97}$$

Stabilized and unreactive radical.

Examples of reactions which can be inhibited include the pyrolyses of saturated hydrocarbons. The rates of these pyrolyses can be reduced, often by a factor of five or so, if the reactions are carried out in the presence of nitric oxide. Hydrocarbon pyrolyses are thought to be initiated by breakage of carbon–carbon bonds. The radicals so formed react in a variety of ways, including attack on hydrocarbon molecules, thereby giving a chain reaction. This process is illustrated below for ethane pyrolysis (termination reactions have not been listed). The removal of alkyl radicals by nitric oxide cuts down the total radical concentration in the system, and hence reduces the rate at which alkane molecules are destroyed by reactions such as (147).

$$CH_3-CH_3 \rightarrow 2CH_3\cdot \tag{144}$$

$$CH_3\cdot + CH_3-CH_3 \rightarrow CH_4 + C_2H_5\cdot \tag{145}$$

$$C_2H_5\cdot \rightarrow C_2H_4 + H\cdot \tag{146}$$

$$H\cdot + CH_3-CH_3 \rightarrow H_2 + C_2H_5\cdot \tag{147}$$

In contrast to alkane pyrolyses, when ethyl and higher alkyl acetates are heated a molecular reaction takes place, giving acetic acid and the appropriate alkene. These reactions are kinetically of first order, and are not inhibited by the addition of cyclohexene.

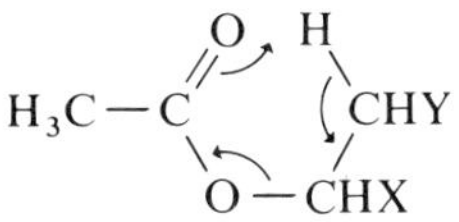

$$H_3C-C\begin{matrix} O \\ \\ O \end{matrix}\begin{matrix} H \\ CHY \\ CHX \end{matrix} \rightarrow CH_3\cdot CO_2H + CHX{=}CHY \tag{148}$$

Arguments based on inhibition must be used with care, since inhibitors for one reaction may act as promoters or initiators for others; for example, nitric oxide inhibits alkane pyrolysis, but initiates the pyrolysis of acetaldehyde. As an example of a misleading result obtained from inhibition experiments, it was found that quinols did not inhibit the autoxidation of trialkyl boranes, and this suggested that these autoxidations are molecular in mechanism (149) rather than proceeding by a chain process (150 and 151) analogous to hydrocarbon autoxidation [reactions (134) and (135)].

$$\begin{matrix} R & O \\ | & || \\ R_2B & O \end{matrix} \rightarrow \begin{matrix} R-O \\ | \\ R_2B-O \end{matrix} \tag{149}$$

$$R\cdot + O_2 \rightarrow R-O-O\cdot \tag{150}$$

$$R-O-O\cdot + R_3B \rightarrow R-O-O-BR_2 + R\cdot \tag{151}$$

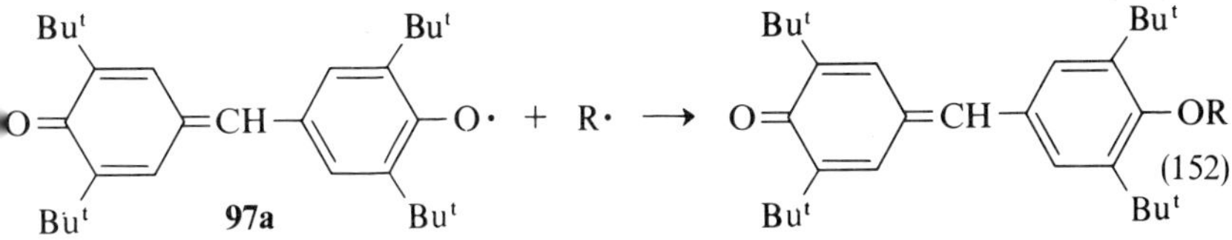

$$\tag{152}$$

However, more powerful inhibitors such as galvinoxyl **97a** will in fact inhibit trialkyl borane autoxidation, and the fact that optically active organoboranes are racemized during autoxidation (compare next chapter, p. 90) also points to a radical rather than a molecular mechanism. With hindsight, we can attribute the failure of quinol to inhibit trialkyl borane autoxidations to the reactivity of the radical **98**. Since this radical will not readily abstract hydrogen from alkanes, the corresponding quinol will be a good inhibitor of alkane autoxidations. However radical **98** appears to be reactive enough to expel an alkyl radical from a trialkyl borane molecule (reaction 153). This will re-start the chain reaction, and hence quinols will not act as inhibitors for trialkyl borane autoxidations.

$$HO-\!\!\!\left\langle\!\!\bigcirc\!\!\right\rangle\!\!-O\cdot \;+\; R_3B \;\longrightarrow\; HO-\!\!\!\left\langle\!\!\bigcirc\!\!\right\rangle\!\!-O-BR_2 \;+\; R\cdot \quad (153)$$

98

Initiation

Free-radical reactions can often be increased in rate by the introduction into the reaction mixture of compounds, termed initiators or promoters, which give rise to radicals on heating or photolysis. Azo-compounds and peroxides are often used as thermal initiators; ketones and other compounds give radicals on photolysis. As mentioned earlier, nitric oxide inhibits the pyrolysis of alkanes, but initiates the pyrolysis of aldehydes, presumably by abstraction of the relatively weakly bound hydrogen atom in the CHO group (reaction 154).

$$CH_3\cdot CHO + NO \;\longrightarrow\; CH_3\cdot CO\cdot + HNO \quad (154)$$

$$CH_3\cdot CO\cdot \;\longrightarrow\; CO + CH_3\cdot \xrightarrow{CH_3\cdot CHO} CH_4 + CH_3\cdot CO\cdot \quad (155)$$

The increase in rate given when a radical initiator is added to a reaction system is not altogether satisfactory as a criterion of the occurrence of a radical mechanism: the non-initiated reaction may be ionic or molecular in mechanism, and the increase in rate when an initiator is added may be due to an entirely different reaction, which does not occur in the absence of the initiator. As an example, hydrogen bromide adds to pure alkenes in the absence of light or radical catalysts to give the alkyl bromide. The rate of this reaction is not reduced by inhibitors, which can correctly be used as evidence that the reaction is not free-radical in mechanism. In fact, the reaction is thought to involve a carbonium-ion intermediate (156).

$$RCH\!=\!CH_2 + HBr \;\longrightarrow\; R\overset{+}{C}H-CH_3 + Br^- \;\longrightarrow\; RCHBr-CH_3 \quad (156)$$

99

The rate of consumption of olefin is greatly increased if free-radical initiators (oxygen, peroxides, etc.) are present: however, this is not because (156) has a radical mechanism, but because an entirely different radical chain reaction (157) is initiated.

$$\text{Br} \cdot + \text{RCH}{=}\text{CH}_2 \;\rightarrow\; \text{R}\overset{\cdot}{\text{C}}\text{H}{-}\text{CH}_2\text{Br} \;\xrightarrow{\text{HBr}}\; \text{RCH}_2\text{CH}_2\text{Br} + \text{Br} \cdot \quad (157)$$
$$\textbf{100}$$

For ethylene and symmetrical internal olefins, these reactions cannot be distinguished by their products, but if a terminal olefin is used the reactions may be distinguished by their products, the secondary bromide **99** from the carbonium-ion reaction, the primary bromide **100** from the radical chain reaction. See also pp. 104–5.

Carbenes

Carbenes are neutral reactive species in which the central carbon atom is covalently bound to only two other groups, and there are two non-bonded electrons which preserve electrical neutrality. These non-bonding electrons may be paired or unpaired, causing certain differences in expected reactivity. Carbenes can be generated in several ways, for example by photolysis of diazo-alkanes, or by treatment of haloforms with strong base.

$$\text{R}{-}\text{CH}{=}\overset{+}{\text{N}}{=}\overset{-}{\text{N}} \;\xrightarrow{h\nu}\; \text{R}{-}\text{CH}{:} \qquad\qquad (158)$$

$$\text{Bu}^t\text{O}^- + \text{HCCl}_3 \;\rightleftharpoons\; \text{Bu}^t\text{OH} + \text{CCl}_3^- \;\rightarrow\; {:}\text{CCl}_2 + \text{Cl}^- \qquad (159)$$

Carbenes are extremely reactive, and resemble radicals in many ways. However, the two most useful diagnostic reactions make use of the possibility of a carbene forming *two* new covalent bonds. Carbenes $:CR_2$ undergo an insertion reaction into $\geqslant C{-}H$ bonds in organic molecules to give the products $\geqslant C{-}CR_2{-}H$. Unfortunately, this reaction does not discriminate between the different $C{-}H$ bonds in an organic molecule, which renders this reaction of limited diagnostic use since a complex mixture of products is usually formed. However, carbenes insert still more readily into $Si{-}H$ linkages, so that a compound such as triethylsilane provides a good trap for carbenes, since only one product is formed (reaction 160).

$$\text{Et}_3\text{Si}{-}\text{H} + \text{R}_2\text{C}{:} \;\rightarrow\; \text{Et}_3\text{Si}{-}\text{CR}_2{-}\text{H} \qquad (160)$$

Triethylsilane is an unsuitable trap if the reaction has been carried out in a basic solution, since it is readily hydrolysed to the silanol Et_3SiOH and hydrogen. In basic solutions it is better to rely on a second method by which carbenes can achieve a tetravalent state, namely by adding to a carbon–carbon double bond to give a cyclopropane derivative.

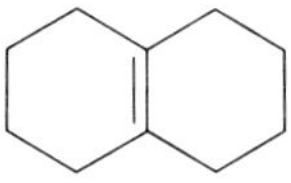

101

Styrene, cyclohexene, and the octahydronaphthalene **101** have frequently been used to trap carbenes, and the reaction may be used as a convenient synthesis of cyclopropane rings.

$$CHCl_3 + Bu^tO^- \rightarrow CCl_2:$$

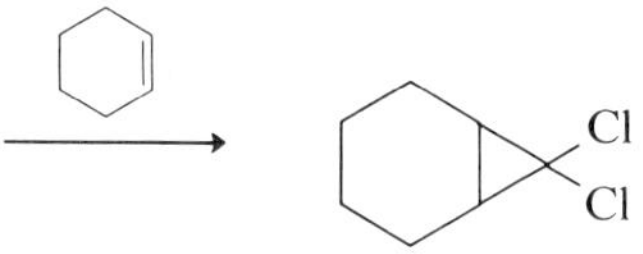

102

(161)

Further evidence that, at least in some cases, genuine free carbenes are formed, and that the reactions are not due to a precursor, has been provided for dichlorocarbene which was prepared in various ways including reaction (159) and the pyrolysis of chloroform which gives $:CCl_2$ and hydrogen chloride. The dichlorocarbene was allowed to compete for a mixture of olefins. For several different pairs of olefins the product ratios were the same no matter how the dichlorocarbene was prepared. This is good evidence that the dichlorocarbene is indeed free—if precursors were involved the different methods of preparation would have given different product ratios.

Benzynes

These reactive intermediates are dehydro-aromatic compounds; benzyne itself, **103**, may be written with a formal triple bond

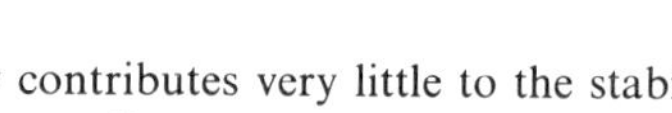

(162)

103

though this probably contributes very little to the stability of the molecule. Evidence for these intermediates, based on unusual substitution patterns, has already been mentioned. Other evidence for benzynes includes the observation of a transient compound of m/e 76 by mass spectrometry when compound **108** is photolysed, and the results of trapping experiments. Benzynes are somewhat analogous to carbenes in being able to form two new bonds to a reagent molecule. A good example is the reaction of benzynes with 1,3-dienes to form six-membered ring compounds: this reaction is analogous to the Diels–Alder reaction with the benzyne playing the role of the activated olefin.

Thus benzyne produced by the action of NH_2^- on PhI (or in other ways) will add across the 9,10-positions of anthracene to give triptycene **104**. Perhaps most commonly used as a trap for benzynes is compound **105** which gives naphthalene compounds with the expulsion of carbon monoxide.

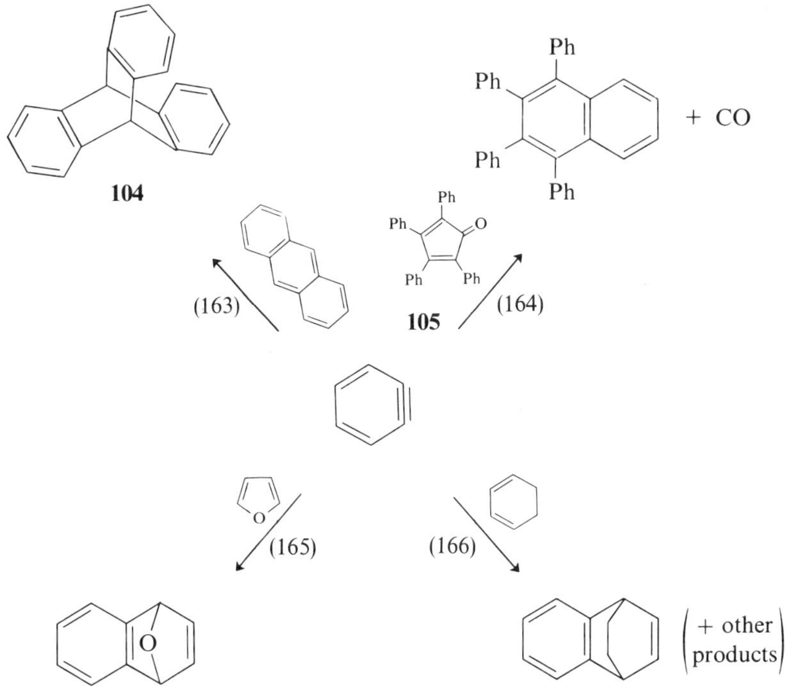

104

Ph
Ph
Ph
Ph
+ CO

Ph
Ph
Ph
Ph
105

(163)

(164)

(165)

(166)

106

107

+ other
products

The existence of 'free' benzynes (rather than precursors which react similarly) has been established in the case of benzyne itself by allowing a mixture of furan and cyclohexa-1,3-diene to compete for it—reactions (165) and (166). Benzyne was prepared in four different ways by the pyrolysis of compounds **108–111**: in each case the ratio k_{165}/k_{166} was almost identical (21.5 ± 0.9), indicating that a common intermediate, free benzyne, is present in all the reactions.

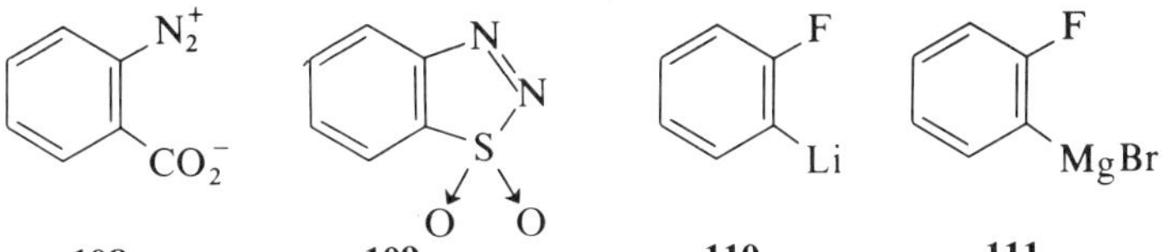

N_2^+
CO_2^-
108

N
N
S
O   O
109

F
Li
110

F
MgBr
111

Tetrahedral intermediates

A number of important reactions of carbonyl compounds (aldehydes, ketones, carboxylic acids, and their derivatives) are thought to involve addition of a reagent to the carbon atom of the carbonyl group to give a tetrahedral intermediate.

$$\ce{\overset{\diagdown}{\underset{\diagup}{}}C=O + X -> \overset{\diagdown}{\underset{\diagup}{}}C\overset{O}{\underset{X}{\diagup}}} \qquad (167)$$

$$\mathbf{112} \qquad\qquad \mathbf{113}$$

(General equation, charges left out)

Intermediates of type **113** are relatively stable when **112** is a ketone, or particularly an aldehyde—here the intermediate **113** or a saturated product derived from it may even be isolated. If **112** is a carboxylic acid, ester, amide, etc. intermediates of type **113** are more reactive, and more difficult to identify and trap. However, there is compelling evidence that even for reactions of carboxylic acid derivatives, structures of type **113** are produced as true intermediates, not just as transition states.

$$\ce{R-C\overset{^{18}O}{\underset{OR'}{}} <=>[HO^-] R-\overset{^{18}O^-}{\underset{\underset{OR'}{|}}{C}}-OH} \qquad (168)$$

$$\mathbf{114} \qquad\qquad \mathbf{115}$$

$$\ce{R-C\overset{^{18}O}{\underset{OR'}{}} <=> R-\overset{^{18}OH}{\underset{\underset{OR'}{|}}{C}}-O^-}$$

$$\mathbf{116} \qquad\qquad \mathbf{117}$$

$$\ce{R-\overset{^{18}O}{C}-OH + OR'^-}$$

$$\mathbf{118}$$

$$\ce{R-C\overset{O}{\underset{OR'}{}} + H+ <=> R-C\overset{OH}{\underset{OR'}{}} <=> R-\overset{OH}{\underset{\underset{OR'}{|}}{C}}-\overset{+}{O}H2 <=> R-\overset{OH}{\underset{\underset{HOR}{|}}{C}}-OH -> R-C\overset{OH}{\underset{OH}{}}} \qquad (169)$$

$$\ce{R-\overset{OH}{\underset{\underset{OR}{|}}{C}}-OH \qquad R-C\overset{O}{\underset{OH}{}} + H+}$$

$$\mathbf{119}$$

The most compelling evidence that tetrahedral intermediates **115** are formed in the ester hydrolysis reaction (168) is provided by the fact that if the original ester is labelled with ^{18}O in the carbonyl group exchange of the ^{18}O with the solvent takes place, so that if the reaction is not carried out to completion the residual ester has less ^{18}O than the original ester. This exchange could not take place if **115** were merely a transition state between **114** and **118**: accordingly the simplest explanation is that **115** is a true intermediate, which can undergo a proton shift to give **117**, which in turn can lose $^{18}OH^-$ to give back the unlabelled ester **116**. Similar arguments apply to the acid-catalysed hydrolysis (169), where again loss of ^{18}O in the carbonyl group during the reaction can be demonstrated.

Intermediates of type **115** or **119** have not been positively identified or trapped in simple esterification or ester hydrolysis reactions. Nonetheless further evidence for these intermediates is available from related reactions. The formation of **120** by the reversible reaction (170) when ethyl trifluoroacetate was treated with sodium ethoxide in di-n-butyl ether was indicated by the reduction in the intensity of the carbonyl stretching band in the infrared.

$$CF_3-C{\overset{O}{\underset{OEt}{}}} + EtO^-Na^+ \rightleftharpoons CF_3-\underset{\underset{OEt}{|}}{\overset{\overset{O^-\quad Na^+}{|}}{C}}-OEt \qquad (170)$$

$$\mathbf{120}$$

Although it has not proved possible to trap tetrahedral intermediates in ester hydrolyses or related reactions by the addition of other molecules to the intermediate, a reactive group in the same molecule may react and thereby give evidence for the intermediate. Thus evidence for the intermediate **122** in the ammonolysis of the ester **121** to the amide **123** (which actually undergoes ring-closure to give **124**) is provided by the formation of **125** in the products by nucleophilic displacement of bromide by the $-O^-$ group in **122**.

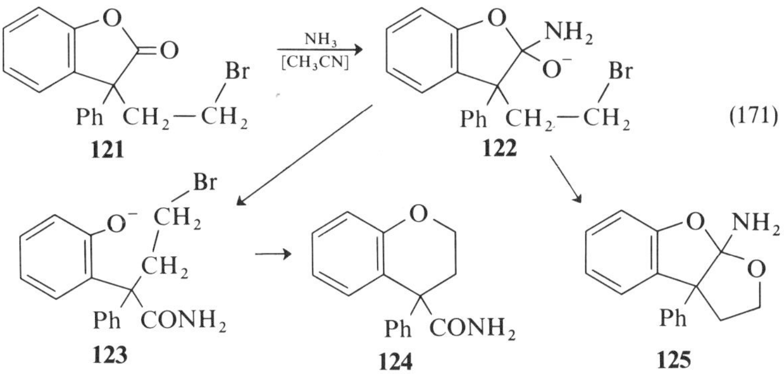

Finally, there is much kinetic evidence in support of tetrahedral intermediates in these reactions. As one example, the base-catalysed hydrolysis of N-methylacetanilide in water at 25°C gives a rate equation which has terms both first- and second-order in the base concentration (compare the self oxidation/reduction of aldehydes considered on page 38). The second-order term in base concentration is extremely difficult to understand except on the basis of the reaction scheme (172) involving a tetrahedral intermediate **127** which in very basic conditions can lose a proton to give the doubly-charged anion **128**. Application of the steady-state treatment to scheme (172) gives a hydrolysis rate of the form

$$-\mathrm{d}[\mathbf{126}]/\mathrm{d}t = k_1[\mathbf{126}][\mathrm{OH}^-] + k_2[\mathbf{126}][\mathrm{OH}^-]^2.$$

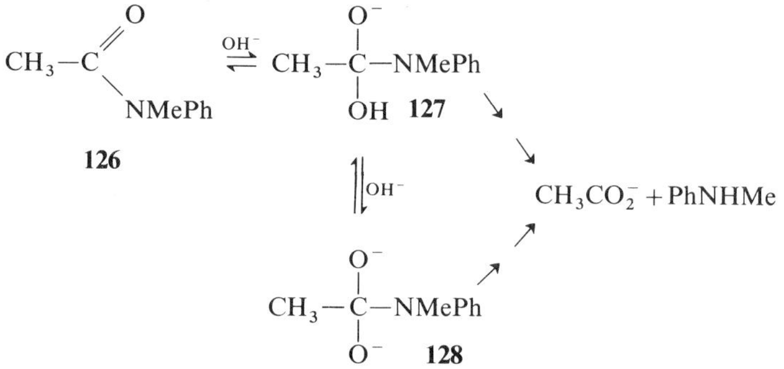

It is easier to obtain evidence for tetrahedral intermediates in reactions which involve aldehydes or ketones, since in some cases stable addition products such as chloral hydrate $Cl_3C\cdot CH(OH)_2$ and cyanohydrins $RCH(OH)CN$ are formed. However, in many reactions of ketones and aldehydes the tetrahedral intermediates which are formed quickly lose water. The condensation of hydroxylamine with acetone is a typical example (reaction 173).

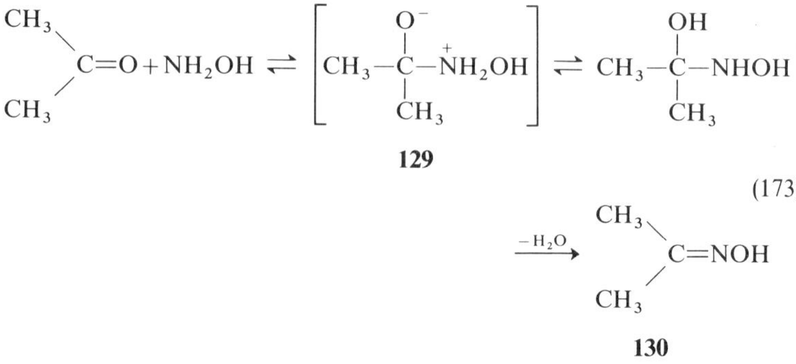

Here the rate of disappearance of acetone, as measured by its u.v. absorption, is greater than the rate of appearance of the oxime **130**. Hence an intermediate must be formed, which is probably **129**. Additional evidence for an intermediate is given by the variation in rate of reaction with pH, which is at a maximum at pH 4·5, falling off in more acid and in more basic solutions. This behaviour implies a reaction which takes place in at least two consecutive steps (and therefore involves an intermediate) since if for example the reaction were a simple acid-catalysed reaction the rate should rise continuously with increasing acidity. The observed behaviour is consistent with mechanism (173). In neutral or basic solutions the equilibrium between the starting materials and **129** is rapidly set up, and the rate-determining step is an acid-catalysed dehydration to the oxime **130**. In acid solutions NH_2OH is protonated to $[NH_3OH]^+$ which will not attack the ketone, and if the solution is sufficiently acidic, the addition of (residual) NH_2OH will become rate-determining, with the acid-catalysed dehydration as a fast subsequent step. Thus in acid solution the rate will be lower than the maximum because of the low concentration of free hydroxylamine, whilst in strongly basic solution the formation of oxime will be slower because of the low concentration of protons.

Acid-base catalysis

A large number of reactions are catalysed by acids or bases. In typical examples, such as the acid-catalysed hydrolysis of esters (reaction 169), the reaction takes place in a number of steps, but many of these only involve transfer of a proton from one molecule to another, or a proton jump within a molecule: processes which require little activation energy (unless they are very endothermic). Hence the overall reaction proceeding in several stages in this manner is often faster than possible competing non-catalysed processes. For example two noncatalysed routes for ester hydrolysis would be:

$$CH_3-C\overset{O}{\underset{OEt}{\diagup}} \rightleftharpoons CH_3-C\overset{O}{\underset{OEt^-}{\diagup}}{}^+ \overset{H_2O}{\longrightarrow} CH_3CO_2H + EtOH \quad (174)$$

$$H_3C-C\overset{O}{\underset{HO\diagdown \quad O-Et}{\diagup}}{}_H \longrightarrow CH_3-\overset{O}{\underset{OH}{C}} + HOEt \quad (175)$$

The direct heterolysis (reaction 174) would involve an enormous activation energy, since there is no particular stabilization of the ions produced. Process (175) is forbidden on symmetry grounds (see Chapter 5, page 97) and may also be expected to have a high activation energy. The acid-catalysed reaction

(169) which has no steps which have particularly high activation energy†, is therefore more feasible than these two more direct processes.

Some reactions are catalysed specifically by H_3O^+ (in water). Other reactions are catalysed additionally by other acids which are present in the system. These two types of behaviour are referred to as specific and general acid catalysis. Similarly, specific and general base catalysis is possible. The occurrence of specific or general catalysis throws light on mechanistic details. For example, a number of acid-catalysed reactions involve addition of a proton to the substrate, followed by further reactions of the adduct.

$$H^+ + X \underset{-1}{\overset{1}{\rightleftharpoons}} HX^+ \overset{2}{\longrightarrow} \text{products} \qquad (176)$$

$$H^+ + X \rightleftharpoons HX^+$$
$$\qquad\qquad\qquad\qquad \searrow \text{products} \qquad\qquad (177)$$
$$HA + X \rightleftharpoons A^- + HX^+ \nearrow$$

If a true prior equilibrium is set up between the substrate and its protonated form HX^+ (reaction 176)—which will be so if both reaction 1 and -1 are very fast compared with 2—then the concentration of HX^+ will be dependent only on the acidity (pH) of the solution, and hence specific acid catalysis will be observed. If however, reaction 2 is comparable with -1 in rate, $[HX^+]$ will not achieve its equilibrium value and the rate at which protons can add to X becomes important. Protons can be added either from H_3O^+, or from other acid molecules present (177): hence terms from all acids present will be present in the kinetic expression, and general acid catalysis takes place. For example, the hydrolysis of ethyl orthoacetate **48** (Chapter 3, reaction 78) is subject to general acid hydrolysis, whereas the rather similar hydrolyses (178, 179) of ethyl orthoformate, **131** and diethylacetal, **133**, are both catalysed specifically by H_3O^+. A possible explanation for this difference is that loss of ethanol from the intermediate cations is easier for **49** than for **132** and **134** because the resultant cation **135** is more stabilized than **136** (methyl substituents stabilize cations) and than **137**, which has only one EtO group to stabilize the positive charge. This makes it reasonable that in these reactions only for ethyl orthoacetate is step (2) comparable in rate with (1) and (-1), hence giving general acid catalysis. It should be remembered that for appropriate compounds, k_2/k_1 or k_2/k_{-1} may have every conceivable value. Thus reactions which are apparently specifically-catalysed, if carried out in buffer of very high acid concentration, might show themselves to be subject to general acid catalysis.

† Transfer of protons to and from the ester molecule should really be regarded as transfer of a proton from H_3O^+ to the molecule and vice versa. Hence a bond is broken and a bond is formed, making the transfer approximately thermoneutral.

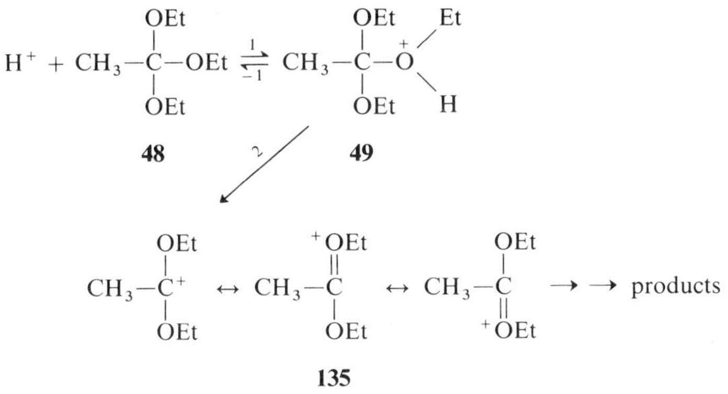

$$H^+ + CH_3-\underset{\underset{OEt}{|}}{\overset{\overset{OEt}{|}}{C}}-OEt \underset{-1}{\overset{1}{\rightleftharpoons}} CH_3-\underset{\underset{OEt}{|}}{\overset{\overset{OEt}{|}}{C}}-\overset{+}{O}\overset{Et}{\underset{H}{\diagdown}}$$

$$\textbf{48} \qquad\qquad \textbf{49}$$

$$CH_3-\underset{\underset{OEt}{|}}{\overset{\overset{OEt}{|}}{C^+}} \leftrightarrow CH_3-\underset{\underset{OEt}{|}}{\overset{\overset{+OEt}{||}}{C}} \leftrightarrow CH_3-\underset{\underset{+OEt}{|}}{\overset{\overset{OEt}{|}}{C}} \rightarrow \rightarrow \text{products} \quad (78)$$

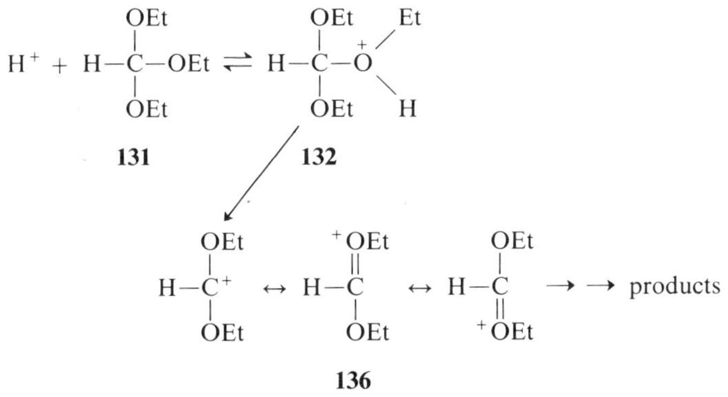

$$\textbf{135}$$

$$H^+ + H-\underset{\underset{OEt}{|}}{\overset{\overset{OEt}{|}}{C}}-OEt \rightleftharpoons H-\underset{\underset{OEt}{|}}{\overset{\overset{OEt}{|}}{C}}-\overset{+}{O}\overset{Et}{\underset{H}{\diagdown}}$$

$$\textbf{131} \qquad\qquad \textbf{132}$$

$$H-\underset{\underset{OEt}{|}}{\overset{\overset{OEt}{|}}{C^+}} \leftrightarrow H-\underset{\underset{OEt}{|}}{\overset{\overset{+OEt}{||}}{C}} \leftrightarrow H-\underset{\underset{+OEt}{|}}{\overset{\overset{OEt}{|}}{C}} \rightarrow \rightarrow \text{products} \quad (178)$$

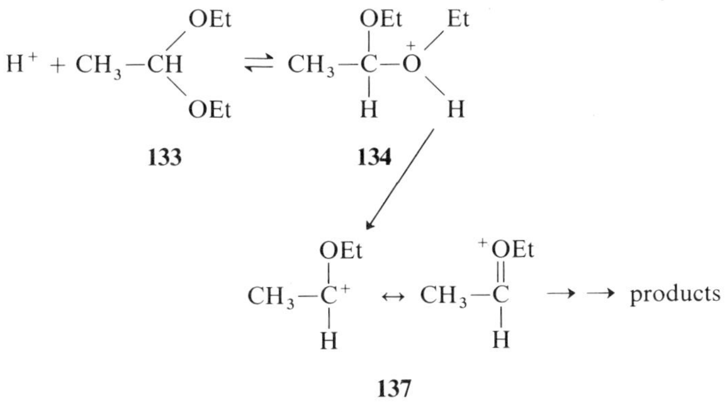

$$\textbf{136}$$

$$H^+ + CH_3-\underset{\underset{OEt}{\diagdown}}{\overset{\overset{OEt}{\diagup}}{CH}} \rightleftharpoons CH_3-\underset{\underset{H}{|}}{\overset{\overset{OEt}{|}}{C}}-\overset{+}{O}\overset{Et}{\underset{H}{\diagdown}}$$

$$\textbf{133} \qquad\qquad \textbf{134}$$

$$CH_3-\underset{\underset{H}{|}}{\overset{\overset{OEt}{|}}{C^+}} \leftrightarrow CH_3-\underset{\underset{H}{|}}{\overset{\overset{+OEt}{||}}{C}} \rightarrow \rightarrow \text{products} \quad (179)$$

$$\textbf{137}$$

Stable molecule intermediates

Sometimes a relatively stable molecule may be formed as a reaction intermediate. The presence of such an intermediate may be indicated by kinetic studies if the initial rate of disappearance of starting materials is greater than the rate of formation of products. If the reaction is followed spectrophotometrically the presence or absence of stable intermediates is often indicated from the changes in the spectrum as the reaction proceeds (Figure 4). Suppose a compound X with u.v. spectrum 1 is converted into Y with u.v. spectrum 5 (see Fig. 4), and that no other absorbing species are present. If X and Y

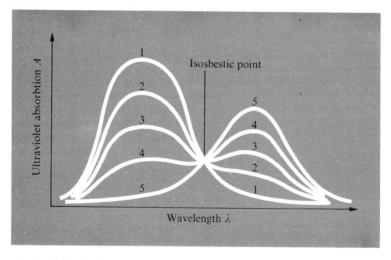

Fɪɢ. 4. The isosbestic point. Ultraviolet absorption curves obtained when a reaction X → Y is followed spectrophotometrically. Curve 1 corresponds to pure **X**, curves 2–4 are of **X** and **Y** in the proportions 75:25, 50:50, 25:75. Curve 5 corresponds to pure **Y**.

have sufficiently different spectra, one or more points will exist where their spectra cross. At these points, called isosbestic points, the two compounds will have equal extinction coefficients and thus at these wavelengths the optical density of the solution will always be the same throughout the reaction. Hence for intermediate times of reaction, spectra of types 2, 3, and 4 will be observed, but each will go through the isosbestic point. Behaviour of the type shown in Fig. 4 is good evidence that a long-lived intermediate is not being formed, whereas spectra which do not show an isosbestic point provide good evidence for the presence of an intermediate.

Isolation or trapping of a stable intermediate may be possible using methods which depend on the actual example. The hydrochloric acid catalysed re-arrangement of *N*-chloroacetanilide, referred to in Chapter 2, gives the same mixture of *o*- and *p*-chloroacetanilide as is produced by direct chlorination of acetanilide, suggesting that free chlorine is formed as an intermediate. This hypothesis was tested in two ways. When *N*,2,4-trichloroacetanilide was heated with hydrochloric acid in the presence of amines such as *p*-bromo-aniline, substantial quantities of 4-bromo-2,6-dichloroaniline were produced, as is to be expected if free chlorine is produced as an intermediate, since *p*-bromoaniline is much more readily chlorinated than 2,4-dichloroacetanilide.

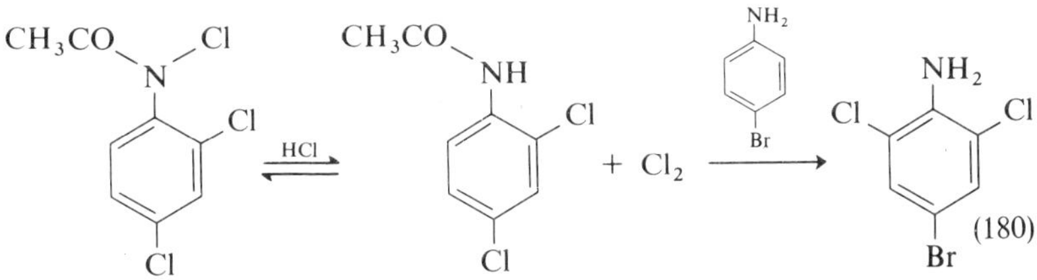

Even more direct and convincing was the demonstration that if nitrogen was swept through the reacting system chlorine was removed from the solution, proving that chlorine is formed during the reaction.

In some cases a relatively small change in conditions (usually a lowering of temperature) allows the isolation of an intermediate which normally reacts too quickly to be detected. Grignard reagents react with oxygen in a well known reaction to give alcohols. It is difficult to write a convincing one-step reaction for this oxidation, and the peroxide **138** seems an obvious intermediate. Such intermediates have been isolated by passing oxygen into Grignard reagents held at $-70°C$.

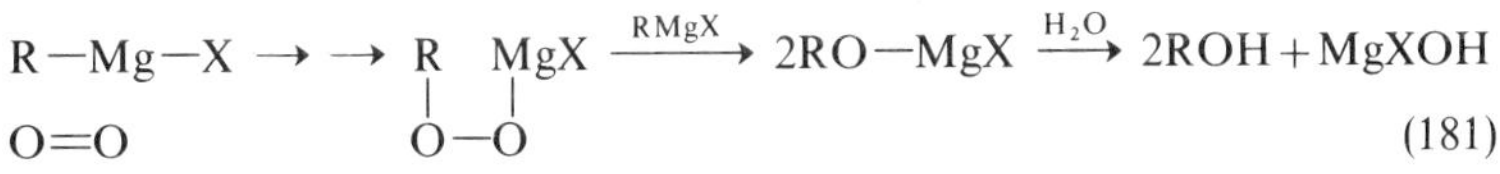

$$R-Mg-X \rightarrow \rightarrow \underset{\underset{O-O}{|\quad|}}{R \quad MgX} \xrightarrow{RMgX} 2RO-MgX \xrightarrow{H_2O} 2ROH + MgXOH$$

$$O=O$$

$$\textbf{138}$$

(181)

Isolated at $-70°$

In the *para* Claisen rearrangement the various labelling experiments suggest that the alkyl group remains attached to the parent molecule, implying an intermediate of the type **139**, which is a diene and as such should undergo the Diels–Alder reaction with a suitable olefin. In fact, addition to maleic an-hydride gives rise to the adduct **140**, which adds still more confidence to the mechanism suggested.

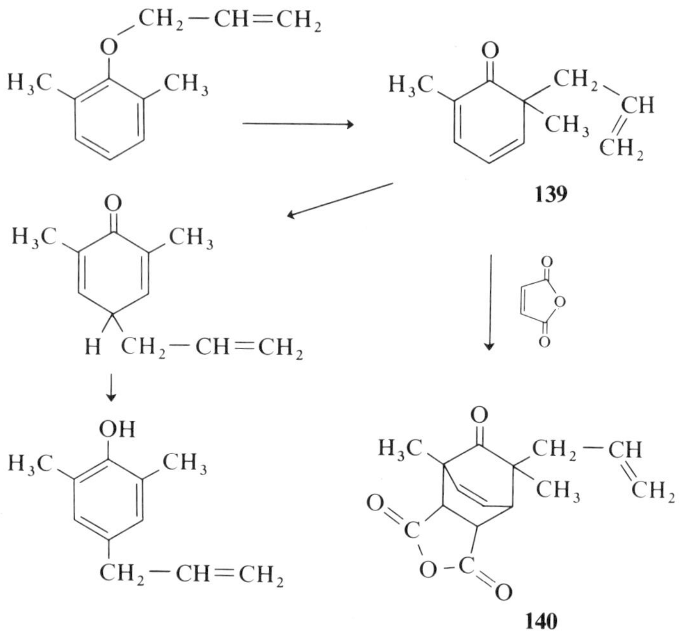

(182)

Conclusion

The occurrence of a particular type of intermediate is normally postulated as a result of product or kinetic studies. The intermediate may then be identified or trapped as outlined above. However, there are two pitfalls to be avoided.

(1) Some of the techniques are extremely sensitive. This is particularly true of e.s.r., CIDNP, and initiation of polymerization. The possibility must always be considered that the 'intermediate' detected by such a technique is the product of a parallel, minor, reaction, and does not feature in the main reaction considered. This possibility is less likely if a trapping technique is used, particularly if the trapping agent diverts all or most of the intermediate away from the usual product of the reaction.

(2) A trapping or inhibiting agent may be insufficiently reactive towards a particular intermediate to divert it from its normal course of reaction, or the agent may react in such a way as to regenerate another reactive intermediate, as can happen in the autoxidation of boranes considered above. Such results may lead erroneously to the conclusion that a particular intermediate or type of intermediate is absent.

PROBLEMS

4.1. The compound $PhCHMe—CHMe·NH_2$ reacts with nitrous acid in acetic acid to give a mixture of $PhCHMe—CHMe·OAc$ (44 per cent), $PhCMe(OAc)—CH_2Me$ (24 per cent) and $PhCH(OAc)—CHMe_2$ (32 per cent) $(Ac=CH_3·CO)$. It is thought that $PhCHMe—CHMe^+$, $PhCHMe—CHMe·$, or $PhCHMe—CHMe^-$ is involved as an intermediate. Which is the most likely?

4.2. Dibenzylmercury, in solution in octane, decomposes thermally to give a quantitative yield of mercury, with bibenzyl as the major organic product. However, traces of 1,2,3-triphenylpropane and toluene are also produced. Suggest a mechanism to account for these observations and further experiments which might provide support for your mechanism.

4.3. Grignard reagents [organomagnesium halides, regarded here for simplicity as having the structure $\overset{\delta-}{R}–\overset{\delta+}{Mg}–Hal$ (**A**) or R^- $^+Mg–Hal$ (**B**)] do not decompose diethyl ether, whereas methyl sodium reacts with diethyl ether to give ethylene, methane and sodium ethoxide. Suggest a mechanism for this reaction. Does the difference in reactivity between methyl sodium and the Grignard reagents throw any light on whether Grignard reagents have the covalent structure **A** or the carbanion structure **B**?

4.4. Primary aliphatic alcohols are oxidized by Cr(VI) in acid solution to give aldehydes and Cr(III). If Mn(II) salts are added to the solution, MnO_2 is precipitated. However, Mn(II) is not oxidized by alcohols, aldehydes, Cr(VI), or Cr(III). Suggest (in general terms) an explanation.

4.5. The compound $PhCHN_2$ is known to give $PhCH:$ on photolysis; if 1-butene is present, an equimolar mixture of *cis*- and *trans*-1-ethyl-2-phenylcyclopropane is formed. The same two products are formed in the proportion 2:1 if $PhCHBr_2$ is treated with lithium in the presence of 1-butene. It is thought that $PhCHBr_2$ reacts with lithium to give $PhCH(Br)Li$, and that this compound either reacts directly with 1-butene to give the observed products, or decomposes to give lithium bromide and the carbene $PhCH:$, which in turn reacts with the 1-butene to give the cyclopropane products. Which mechanism is the more likely?

5. Stereochemistry

In Chapter 1 it was explained that we could not hope to achieve an exact and complete picture of the course of a chemical reaction. However, stereochemical studies allow us further insight into the details of a reaction mechanism by providing information about the direction of approach of a reagent or of the movement of groups during a reaction.

By a paradox, much of our knowledge of stereochemistry is derived from ideas of mechanism, as will be explained later.

Basically, two types of stereo-isomerism exist: optical and geometrical. Useful mechanistic information can be obtained by studying reactions involving molecules which can exhibit one or the other form of stereo-isomerism.

Substitution at tetrahedral centres and optical isomerism

Organic compounds with four different groups attached to a central carbon atom exist in two isomeric forms. These optical isomers have the same chemical reactivity† but differ in their optical properties. In particular they will have equal and opposite optical rotations (solutions of the same concentration will rotate a beam of plane-polarized light by equal and opposite amounts).

In a substitution reaction in which one group attached to the central carbon atom is replaced by a new group, there are two stereochemical possibilities. Either the new group takes the same position relative to the other substituents as the departing group (reaction 183—a process which is referred to as involving retention of configuration;—or the opposite optical isomer is formed (reaction 184). At this stage we will not consider ways in which this might come about, but it should be noted that the position of at least one of the non-reacting groups must change relative to the other groups. This process is called inversion of configuration for reasons which will be detailed later in the chapter. At this stage it is sufficient to note that compound **143** will have the opposite optical rotation to **142**, the compound which would be formed by substitution with retention of configuration. Reactions (183) and (184) are the only two possibilities at the molecular level. On the macroscopic scale, it may be found that reaction occurs entirely by (183) or entirely by (184); alternatively both reactions may take place concurrently, in which case total or partial loss of optical activity or *racemization* occurs.

Demonstration that a particular reaction involves retention or inversion of configuration is not a trivial exercise since (at least until very recently) it has not been possible to relate the sign of rotation of an optical isomer with its

† Except in reactions with asymmetric reagents.

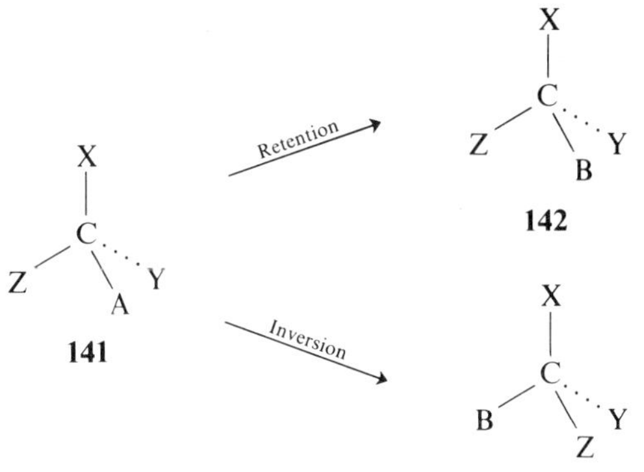

$$
\underset{\textbf{141}}{\underset{A}{\overset{X}{\underset{Z}{\bigwedge}}}Y}
\quad
\begin{array}{c} \xrightarrow{\text{Retention}} \\ \\ \xrightarrow{\text{Inversion}} \end{array}
\quad
\underset{\textbf{142}}{\underset{B}{\overset{X}{\underset{Z}{\bigwedge}}}Y}
\quad (183)
$$

$$
\underset{\textbf{143}}{\underset{Z}{\overset{X}{\underset{B}{\bigwedge}}}Y}
\quad (184)
$$

absolute configuration (e.g. is it **142** or **143**?). Further, it cannot even be assumed that compounds with the same relative configuration (i.e. which show the relationship that **141** has to **142**) will have the same sign of optical rotation. This can be demonstrated by considering the three reactions (185) in which only one group is replaced at a time, always with retention of relative stereochemistry. The net result of the three reactions is to get back to the optical isomer of the original compound. Clearly, either one or all three of the reactions used must have resulted in a change of sign of rotation.

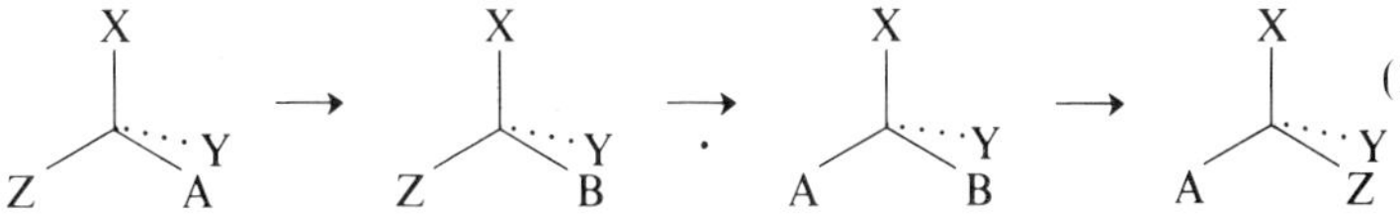

$$
\underset{Z\quad A}{\overset{X}{\bigwedge}}Y
\rightarrow
\underset{Z\quad B}{\overset{X}{\bigwedge}}Y
\cdot
\underset{A\quad B}{\overset{X}{\bigwedge}}Y
\rightarrow
\underset{A\quad Z}{\overset{X}{\bigwedge}}Y
\quad (185)
$$

To determine the stereochemical course of a substitution reaction, it is necessary to know the relative configurations of the starting materials and the products. Various methods for determination of relative configuration are described in the next three sections.

The Walden inversion

In 1896, Walden carried out the reactions shown in scheme (186).

$$
\underset{(+)\text{-chloro-succinic acid}}{HO_2C\cdot CH_2\cdot CHCl\cdot CO_2H} \xrightarrow{\text{`AgOH'}} \underset{(+)\text{-malic acid}}{HO_2C\cdot CH_2\cdot CHOH\cdot CO_2H}
$$

$$
PCl_5 \uparrow \quad \downarrow \overset{KOH,}{_{H_2O}} \qquad\qquad \overset{KOH,}{_{H_2O}} \uparrow \quad \downarrow PCl_5 \qquad (186)
$$

$$
\underset{(-)\text{-malic acid}}{HO_2C\cdot CH_2\cdot CHOH\cdot CO_2H} \xleftarrow{\text{`AgOH'}} \underset{(-)\text{-chloro-succinic acid}}{HO_2C\cdot CH_2\cdot CHCl\cdot CO_2H}
$$

Clearly to account for these results, in which a molecule is converted into its mirror image in two stages, either the reactions shown vertically in the diagram involve retention and the horizontal reactions involve inversion, or vice versa. How can these possibilities be distinguished?

In 1923 Phillips converted $(+)$-benzylmethylcarbinol **144** into the $(+)$ and $(-)$ forms of its ethyl ether by the two reaction sequences shown. Four individual reactions are involved: to account for the stereochemical result either one or three of these must involve inversion of configuration, and the remainder retention. Three of the reactions (187, 189, and 190) do not involve attack at the asymmetric centre at all and it is therefore almost certain that the configuration of the optically active carbon atom is unaltered in these processes. Thus the last remaining reaction (188) which is a bimolecular nucleophilic substitution (S_N2) reaction by an ethanol molecule at the asymmetric carbon atom must be accompanied by effectively 100 per cent inversion of configuration. Further work of this kind established that in a number of other cases of nucleophilic substitution at carbon, inversion of configuration took place in the same way†.

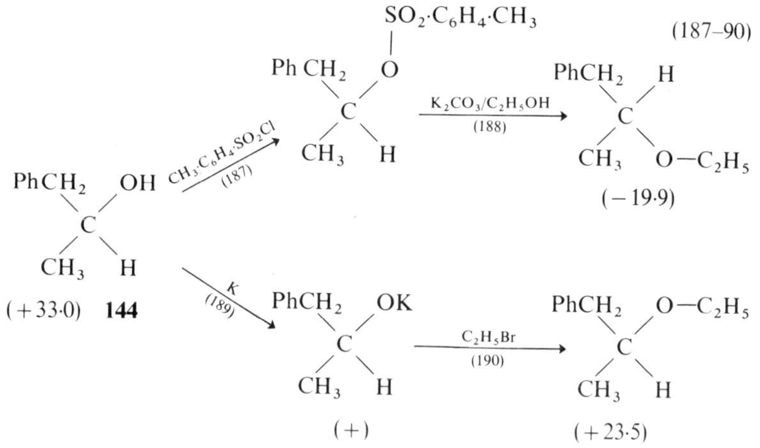

The connexion between bimolecular nucleophilic substitution and inversion of configuration was put on an even firmer basis by Hughes and co-workers in 1935. They studied in separate experiments the racemization of 2-iodo-octane by iodide ions, and the incorporation by 2-iodo-octane of radioactive

† The symbol ⟶⟶ is used to denote a reaction which occurs with inversion of configuration.

iodine from radioactive iodide ions. Both these reactions were of second order, first order with respect to both 2-iodo-octane and iodide ions. The radioactive iodine incorporation experiments established the rate at which the substitution reaction (191) took place.

$$^{128}I^- + R - I \rightarrow \ ^{128}I - R + I^- \tag{191}$$

The stereochemical consequences were shown by the racemization experiments. If each individual exchange reaction took place with retention of configuration, no loss of optical activity would take place. If there were no stereochemical preference in the substitution reaction, the rate of incorporation of iodine would be the same as the rate of loss of optical activity. However, if inversion of configuration took place at each substitution, the rate of racemization would be twice that of incorporation of radioactive iodine, since for example a single substitution on a molecule of the (+)-iodide would give one molecule of the (−)-iodide, which would cancel out the rotation due to another molecule of (+)-iodide.

$$^{128}I^- + R - I \rightarrow \ ^{128}I - R + I^- \tag{192}$$
$$(+) \qquad \qquad (-)$$

Within experimental error, it was established that the rate constant for racemization was twice that for radioactive iodine exchange, and hence each and every individual substitution reaction corresponded to inversion of configuration.

Similar results were obtained for the reaction of bromide ion with optically active 1-phenylethyl bromide. Thus for bimolecular nucleophilic reactions in which the entering and leaving group are the same (2-iodo-octane and 1-phenylethyl bromide), or involve the same element (oxygen) attached to the asymmetric carbon (reaction 188), inversion of configuration was always found to take place. On the basis of these results, it was *assumed* that inversion of configuration would take place even if the nucleophile and leaving group were different. A very great number of structural correlations were made in this way, and no conflicting results for asymmetric carbon compounds have ever been obtained, provided that the nucleophilic substitution reactions can be shown to be bimolecular. As an example of the structural correlations achieved in this way, D-glyceraldehyde was shown to have the opposite relative configuration to the naturally occurring amino-acid (+)-alanine. Since all the common naturally occurring α-amino-acids can be shown to have the same configuration, they all belong to the L-series of compounds.

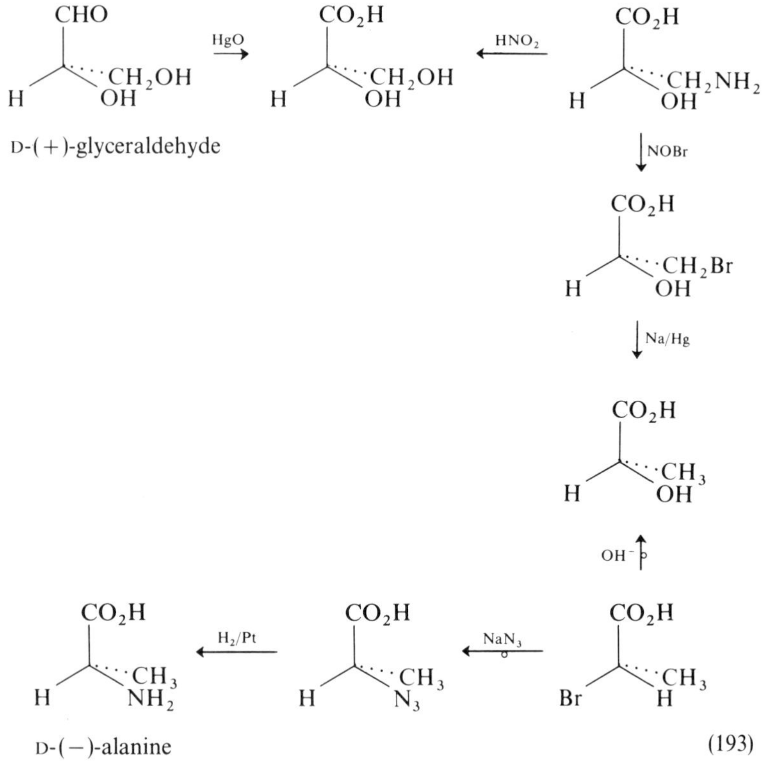

D-(+)-glyceraldehyde

D-(−)-alanine   (193)

The pseudo-racemate method

Although many physical properties (in particular melting points) of optical isomers are the same, being different compounds they will depress each other's melting points. The equimolar mixture of the two isomers may either be a simple eutectic (racemic) mixture, or it may in some cases form a racemic compound. To take an example from organosilicon chemistry, suppose we have two different optically active compounds, such as PhNpMeSiH and PhNpMeSiF (Np = α-naphthyl), where the differences in structure are relatively small. The two compounds of the same configuration will often form a solid solution, in which the lattice sites are filled at random with either of the two molecules, depending on how much of each is present. If the two compounds have opposite configurations no such solid solution will in general be possible, and either a simple eutectic mixture or a 'pseudo-racemic' compound will be formed. These possibilities can be distinguished by melting-point curves, or more usually by X-ray diffraction. By this method it was established for example, that $(+)$-Ph Np MeSiCl, $(-)$-Ph Np MeSiF, and $(-)$-Ph Np MeSiH have the same relative configuration, which establishes the stereochemical course of various interconversions of optically active silicon compounds such as those shown in scheme (194). It has become apparent that the stereochemistry of substitution at silicon is not so simple as at carbon, since for example bimolecular nucleophilic substitutions on silicon sometimes occur with inversion of configuration (194b) and sometimes with retention, as in (195). [If this reaction occurred with inversion, one molecule of $(+)$ and one of $(-)$ product would be produced in each individual encounter, thereby giving an optically inactive (racemic) product.]

$$\text{Ph Np MeSiH} \xrightarrow[\text{(a)}]{\text{Cl}_2/\text{CCl}_4} \text{Ph Np MeSiCl} \xrightarrow[\text{(b)}]{\text{LiAlH}_4} \text{Ph Np MeSiH} \qquad (194)$$

$$(+32) \qquad \text{Retention} \quad (-6\cdot3) \qquad \text{Inversion} \quad (-32)$$

$$\text{Ph Np MeSi}-\text{O}-\text{SiMe Np Ph} \xrightarrow{\text{KOH, xylene}} 2\text{Ph Np MeSiOK} \qquad (195)$$

$$(+9\cdot8) \qquad\qquad\qquad\qquad (-70)$$

Other methods

Another important method of determination of relative configuration is optical rotatory dispersion (the variation of optical rotation with wavelength of light used). As has already been pointed out, the sign of rotation of an optically active compound is not a reliable guide to its stereochemistry. If the optical rotation is measured as a function of wavelength, a curve of one of the types shown in Fig. 5 is produced. If the compound has an absorption band in the visible or ultraviolet caused by a chromophore which is near to the asymmetric centre, a curve of type (a) is found. In general, as the wavelength becomes shorter, the optical rotation changes at first slowly, then more rapidly up to a peak value (positive or negative), then falls rapidly and crosses the zero level at a wavelength corresponding approximately to the λ_{max} for the chromophore. This type of behaviour is called the Cotton effect, and for the two compounds illustrated the effect is defined as positive. (A negative Cotton effect would be the mirror image of (a) with respect to the horizontal axis.) If the absorption maximum occurs at too short a wavelength, a *plain curve* (as shown in Fig. 5b) is obtained. It may be assumed that both the curves shown would exhibit positive Cotton effects if optical rotations could have been measured at shorter wavelengths. The sign of the Cotton effect is a very much better guide to relative stereochemistry than the optical rotation at a particular wavelength. For example, all the naturally occurring α-amino-acids, $RCH(NH_2)\cdot CO_2H$, provided they have no chromophore other than the carbonyl group, show a positive Cotton effect and therefore probably have the same relative configuration in spite of the fact that their optical rotations

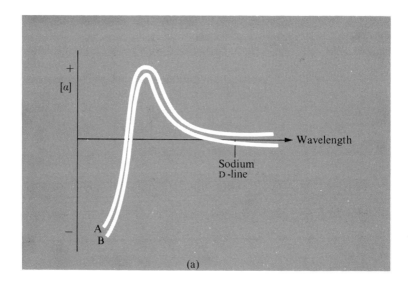

(a)

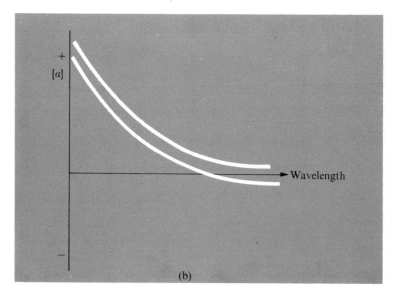

(b)

FIG. 5. Optical rotatory dispersion. (a) O.R.D. curves for two compounds **A** and **B**. Both curves show a positive Cotton effect, but the two compounds have optical rotations of opposite signs at the sodium D-line. (b) Plain curves, i.e. curves without a Cotton effect in the accessible region of the ultraviolet.

at, say, the sodium D-line may be of opposite signs. A very large number of correlations of relative configurations of other natural products, particularly of steroids, have been made in this way.

Relatively recently, it has become possible by a modification of the normal technique of X-ray structural analysis to assign absolute configurations to a number of compounds. This has been done only for a limited number of compounds, but it has allowed a check to be made on the relative configurations of various classes of compounds which have been established by other methods.

Inversion reactions at carbon

The inversion of configuration which accompanies S_N2 reactions implies attack of the nucleophile from the opposite side of the molecule to the leaving group, leading to the transition state **146**, and thence to the product **147** which has the opposite configuration to **145**. Attack from this direction minimizes repulsion forces between the incoming and outgoing groups which are often bulky, and allows formation of a new bond as the old one is broken. In the example shown in scheme (196), both incoming and outgoing groups have a

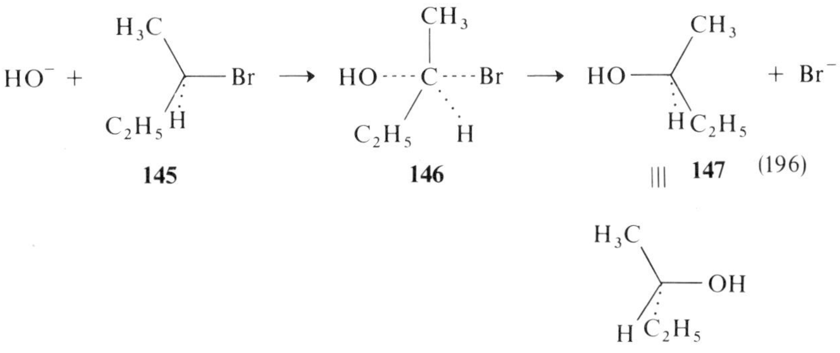

$$C_6H_{13}\cdot\overset{+}{C}HMeSMe_2 + OH^- \;\dashrightarrow\; C_6H_{13}\cdot CH(OH)Me + Me_2S \quad (197)$$

negative charge. However, repulsion between like charges cannot be the reason for the direction of attack, since when the positively charged sulphonium salts are attacked by hydroxide ion, inversion of configuration still takes place, even though the entering and leaving groups have opposite charges (reaction 197).

This picture of the S_N2 mechanism is supported by the fact that if the structure of the molecule is such that the back-side attack cannot take place, as in the bicyclic compound **148**, nucleophilic substitutions do not occur at all readily. The t-butyl substituent in neopentyl bromide **149** is sufficiently bulky to make attack at the CH_2 group by a nucleophile difficult.

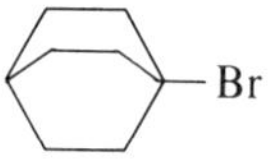

$$\text{148} \qquad \text{149}$$

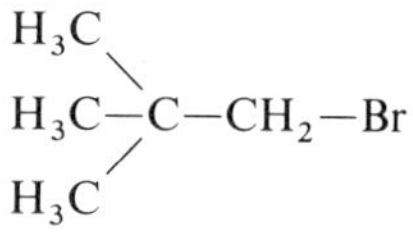

In general, primary compounds undergo S_N2 substitution more readily than do secondary compounds, and S_N2 reactions on tertiary compounds go with great difficulty if at all. In the S_N2 transition state **146**, the central carbon atom has increased its coordination number from four to five, and steric compression between substituents will be more severe than in the ground state. Alkyl groups take up more space than hydrogen atoms, hence this steric compression will be more severe in the order tertiary > secondary > primary compounds.

Substitution with retention

This is much more rarely encountered than inversion in carbon systems. However, two circumstances will favour attack at the front of the molecule.

(1) If the attacking reagent is electrophilic (i.e. electron deficient) repulsion between the incoming and outgoing groups will not be so severe. Electrophiles normally have an empty orbital. This can overlap with the $C-X$ bond in **150** giving a three-centre orbital with some bonding character.

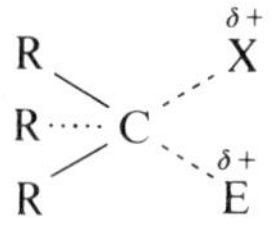

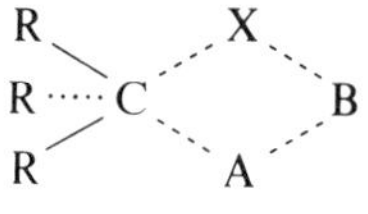

(2) If the substitution is essentially a molecular 4-centre reaction with the transition state **151**, retention of configuration will be expected.

It is often hard to distinguish between possibilities (1) and (2).

Examples of retention of configuration at carbon include a number of reactions of organometallic compounds, such as reaction (198).

It may be noted that for elements beyond the first row of the periodic table, reactions involving retention of configuration are more common. An example involving retention of configuration at silicon was given earlier in the chapter (reaction 194).

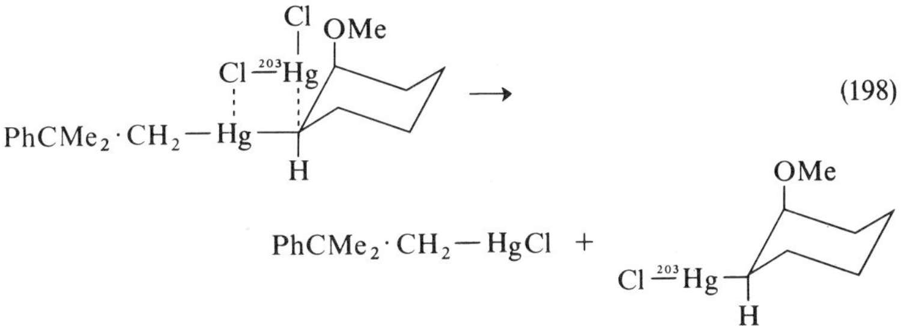

$$\text{PhCMe}_2 \cdot \text{CH}_2 - \text{Hg} \underset{\substack{| \\ \text{H}}}{\overset{\substack{\text{Cl} \\ | \\ \text{Cl} \overset{203}{-} \text{Hg}}}{\diagup}} \overset{\text{OMe}}{\diagdown} \longrightarrow \tag{198}$$

$$\text{PhCMe}_2 \cdot \text{CH}_2 - \text{HgCl} \; + \; \text{Cl} \overset{203}{-} \text{Hg} \underset{\substack{| \\ \text{H}}}{\diagup} \overset{\text{OMe}}{\diagdown}$$

Reactions which involve carbonium ions, radicals, or carbanions as intermediates
Carbonium ions

S_N1 reactions of alkyl halides are thought from kinetic and other evidence to involve carbonium ions. If an optically active halide $RR'R''C-Hal$ reacts in this way the product is usually more or less racemized, which suggests that the free carbonium ion intermediate is planar or nearly so. Supporting evidence for the preferred planarity of the carbonium ion intermediate comes from studies of the reactivities of bridged-ring compounds where the possible intermediate carbonium ion could not become planar because of ring strain;

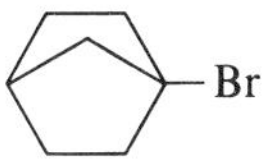

152 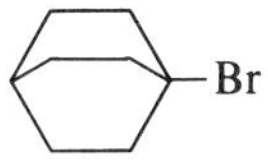   **153** 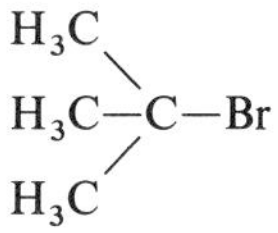   **154**

for example, compound **153** undergoes hydrolysis in aqueous ethanol at 25°C at about a millionth of the rate of the similar reaction of **154**. Compound **152**, in which the formation of a planar carbonium ion would be even more difficult, reacts about ten million times more slowly than **153**.

In actual fact, reactions which involve carbonium ion intermediates do not always give completely racemic products—where the overall result is partial inversion of configuration it is often assumed that although a free carbonium ion has been formed, the departing group (X in **155**) will partially protect that side of the molecule from attack: the incoming group (Y) will therefore preferentially attack from the opposite side and partial inversion will occur.

Incoming group attacks preferentially from this side.	$\text{Y} \quad \underset{\diagup \diagdown}{\overset{|}{\text{C}^+}} \quad \text{X}$	Departing group shields this face of the carbonium ion.

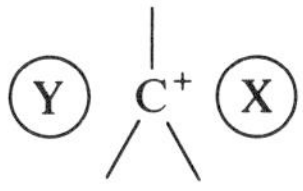

155

Some nucleophilic displacements at carbon occur with overall partial retention of configuration: an example is the reaction of chlorosuccinic acid with silver hydroxide (reaction 186). In a related reaction, the hydrolysis of **156**, inversion of configuration was observed when a high concentration of alkali was used and a bimolecular S_N2 reaction occurred. When more dilute alkali was used the reaction was unimolecular and gave predominant retention of configuration in the product. It is thought that in the carbonium-ion intermediate **157**, there is an interaction between the carboxylate group and the carbonium ion centre which both stabilizes the ion and tends to prevent an attacking nucleophile from reacting at this side of the molecule. The approach of nucleophile is therefore preferentially from the same side as the departing group and predominant retention of configuration occurs.

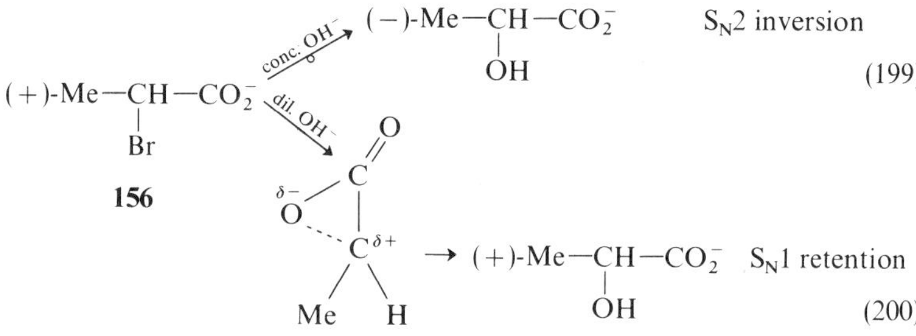

$$(-)\text{-Me}-\underset{\underset{\text{OH}}{|}}{\text{CH}}-\text{CO}_2^- \qquad S_N2 \text{ inversion} \tag{199}$$

$$(+)\text{-Me}-\underset{\underset{\text{Br}}{|}}{\text{CH}}-\text{CO}_2^-$$

156

157

$$\rightarrow (+)\text{-Me}-\underset{\underset{\text{OH}}{|}}{\text{CH}}-\text{CO}_2^- \qquad S_N1 \text{ retention} \tag{200}$$

Radicals

When substitution takes place at an asymmetric carbon centre by a free-radical reaction, the product is optically inactive. Chlorination of optically active 1-chloro-2-methylbutane **158** at the optically active centre gives an inactive product **160**.

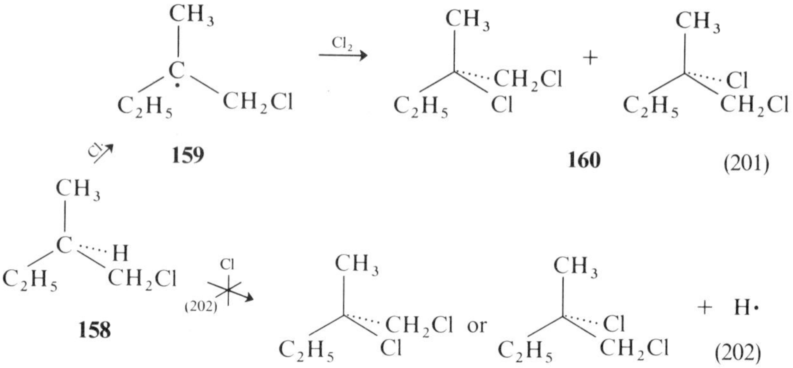

159   **160**   (201)

158

$$\text{161} \qquad \text{162} \qquad + \text{H·} \tag{202}$$

This suggests that the intermediate radical **159** is formed which is effectively planar and thus gives the racemic product. If substitution had taken place directly by the chlorine atom (reaction 202) a stereochemical preference *either* for retention (product **162**) or for inversion (product **161**) would have been expected. Support for the preferred planar structure for aliphatic radicals comes from the e.s.r. spectrum of $^{13}CH_3$·, and the infrared spectrum of the methyl radical. Although simple alkyl radicals have a preferred planar structure, it turns out that halogenated radicals such as CF_3· are *not* planar, and neither are cyclopropyl radicals, which are thereby enabled to undergo radical reactions without complete racemization. The stereochemical preference of alkyl radicals for a planar structure appears to be slight. In contrast to the carbonium ions where bridgehead bicyclic compounds do not readily undergo S_N1 reactions, radical substitution at such bridgeheads or the production of radicals by other means at these positions does not appear to be exceptionally difficult.

Carbanions

Ketones are believed to react with bases to give carbanions which can undergo other reactions or pick up a proton from the solvent to give back the (racemic) ketone [reaction (41), page 19]. Since the rate of racemization of the ketone is the same as the rate of various base-catalysed reactions of the ketone, this is evidence either that the carbanion is planar, or, if it is pyramidal, that it inverts rapidly. However, as with carbonium ions, some reactions thought

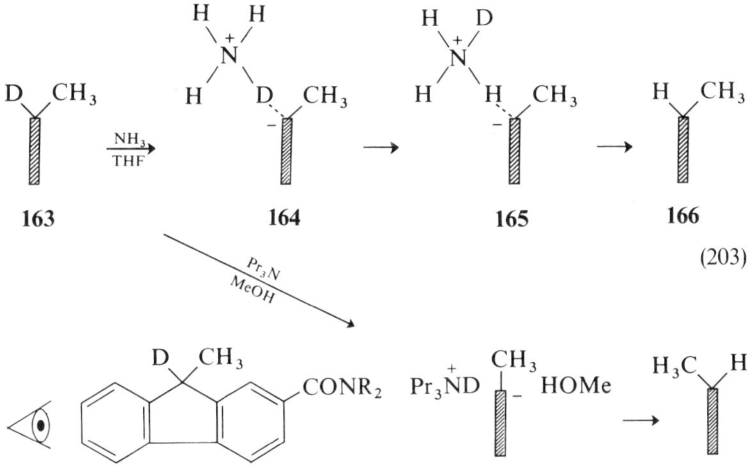

molecule viewed from end in reaction scheme.

(203)

(204)

to involve carbanions display predominant retention and some predominant inversion. The optically active fluorene **163** can display both kinds of behaviour. With ammonia in relatively non-polar solvents, predominant retention of configuration takes place (203). It is thought that an ion pair is formed in which there is considerable interaction between the positive and negative ions, so that it is more likely that the ion pair will return to **163** than dissociate into completely free ions. However, rotation of the ammonium ion can occur within the ion pair so that the recovered fluorenone **166** has hydrogen rather than deuterium, but with predominant retention of configuration. On the other hand, in more polar, ionizing solvents such as methanol, the protonated base molecule will have a greater tendency to escape from the anion **167**. A proton from a solvent molecule is more likely to be abstracted by **167** on the opposite side to the departing $Pr_3\overset{+}{N}H$ ion, thus giving a product **168** with predominant inversion (reaction 204).

Carbonium-ion rearrangements

One of the features which distinguish carbonium ions from radicals and carbanions is their ease of rearrangement by a shift of a group from an adjacent atom to the carbonium ion site. Provided that some interaction can take place

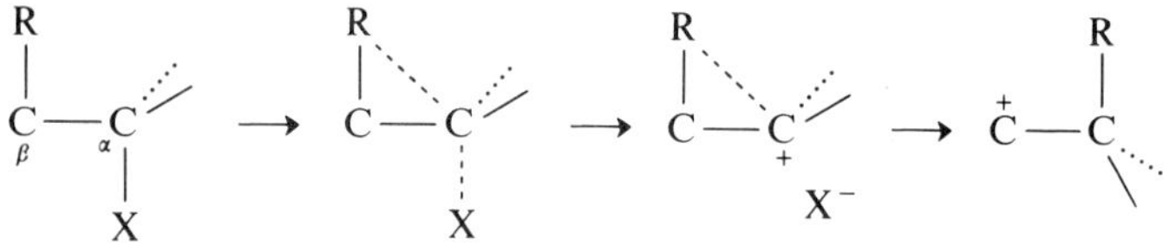

$$(205)$$

169   **170**

between the migrating group and the α-carbon as the group X leaves as X^-, from the point of view of C_α the reaction can be seen as an internal 'S_N2 type' reaction, and the configuration at C_α will be inverted. An interaction of the type shown in **170** is only possible if the four atoms $R-C-C-X$ lie approximately in a plane in the zig-zag (*anti*) conformation shown: in practice only groups which can take up approximately this conformation migrate easily. For example, the diol **171** rearranges to the ketone **172** about six times faster

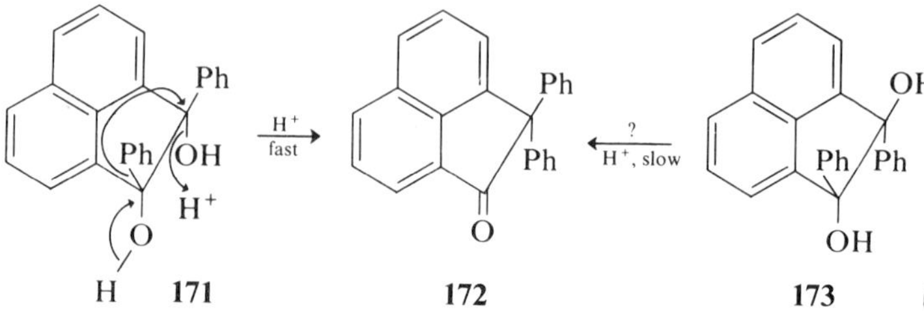

H   171   **172**   **173** (206)

than **173**. In **171**, as the protonated hydroxyl group leaves the α-carbon atom, the β-phenyl group, which is *trans* to the leaving group, migrates to the α-carbon atom. In **173** neither phenyl group is in the favourable position *trans* to the departing hydroxyl group: the rearrangement is very much slower, and it is likely that the slow rearrangement observed is really due to a prior isomerization of **173** to **171**.

If the migrating group in a carbonium-ion rearrangement (R in reaction 205) has an asymmetric carbon atom directly attached to the β-position of **169**, then configuration is retained during the reaction. From the point of view of R the reaction can be seen as an electrophilic substitution by C_α, with displacement of C_β. The 'entering' and 'leaving' groups are on the same side of the asymmetric carbon atom and thus retention takes place.

Additions to multiple bonds and eliminations

Two questions may be asked about reactions involving addition to multiple bonds (reaction 207), or the reverse reactions, termed eliminations:

$$X-Y + \overset{\diagdown}{\underset{\diagup}{C}}=\overset{\diagup}{\underset{\diagdown}{C}} \rightarrow -\overset{|}{\underset{|}{C}}-\overset{|}{\underset{|}{C}}- \tag{207}$$
$$ X Y$$

(1) Are both new bonds formed simultaneously? (2) From what direction(s) does the reagent(s) attack the multiple bond? By using suitably substituted olefins, it may be determined whether the direction of addition of $X-Y$ is *cis* (i.e. both new bonds formed by attack from one side of the molecule), *trans* (one new bond from one side and one from the other), or indiscriminate. We will consider the possibilities and the light they throw on mechanism separately.

Trans (and indiscriminate) addition

If *trans* or indiscriminate addition takes place, the short answer to question (1) above as to whether the two new bonds are formed simultaneously is no. The transition state for *trans* addition must involve attack of X from one side of the π bond and Y from the other, and from these positions they cannot still maintain any significant degree of bonding to each other. Thus for reactions which show *trans* addition a molecular four-centre mechanism (208) must be ruled out.

$$\overset{\diagdown}{\underset{\diagup}{C}}=\overset{\diagup}{\underset{\diagdown}{C}} \rightarrow -\overset{|}{\underset{|}{C}}-\overset{|}{\underset{|}{C}}- \tag{208}$$
$$ X-Y X Y$$

In practice, a very great number of *trans* addition and elimination reactions are known. For example, addition of bromine to maleic acid **174** gives the

($\pm$)-dibromo-acid **175** which can be resolved, whereas addition of bromine to fumaric acid **176** gives the non-resolvable *meso*-acid **177**.

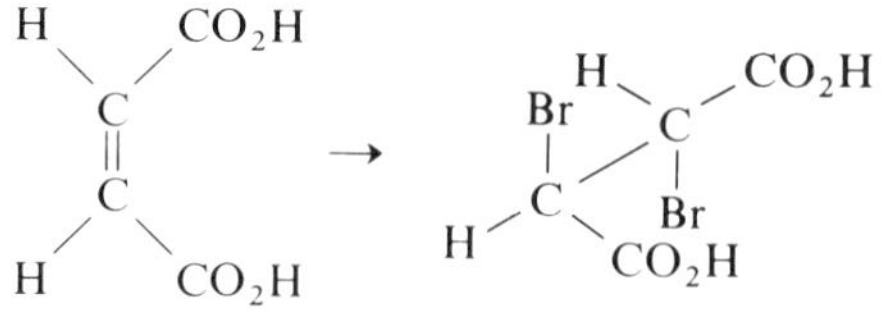

maleic acid
174

($\pm$), resolvable
175

(209)

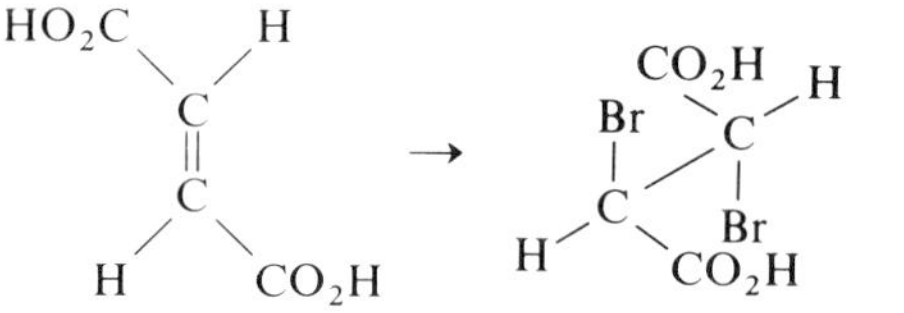

fumaric acid
176

meso, non-resolvable
177

(210)

Many additions to unsaturated cyclic compounds and to acetylenes have also been found to be *trans* in stereochemistry. *Trans* additions often occur in reactions which are believed (from evidence about intermediates formed) to involve initial addition of an electrophile or a radical to one end of the double bond. This raises another difficulty. If, for example, the carbonium ion **178** is formed as an intermediate in the addition of bromine to maleic acid (reaction 209) we might expect indiscriminate stereochemistry rather than *trans* addition since rotation about the single carbon–carbon bond (a) in **178** should be rapid, giving an equal opportunity for Br$^-$ to add from either direction. This difficulty is overcome by assuming that there is an interaction between the bromine atom and the carbonium-ion site, either as a bonding contribution, as pictured in **179**, or giving a symmetrical intermediate **180**. In either case, free rotation about the C—C bond will be reduced or prevented, and the cation will be able to pick up Br$^-$ from the opposite face, both for steric reasons, and also (analogously to S$_N$2 reactions) because bonding can be maintained best (and therefore the energy of the transition state kept as low as possible) if reaction occurs in this way.

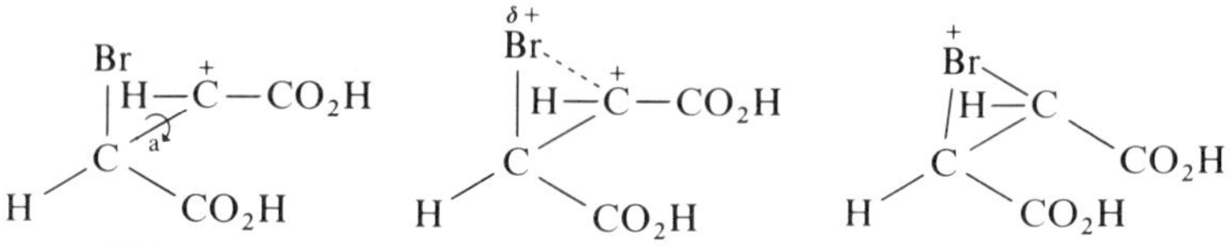

178

179

180

Eliminations are normally carried out under rather different conditions from the corresponding addition reactions: typical reactions include the elimination of hydrogen halides from alkyl halides by strong bases (211), and the elimination of bromine from dibromides, promoted by iodide ion (212).

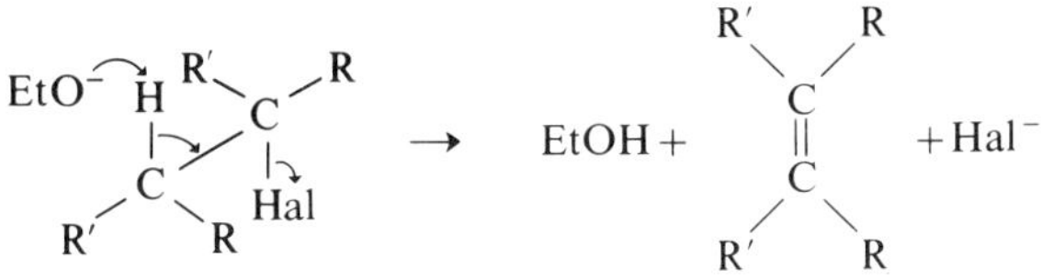

$$EtOH + \quad \underset{R'\quad R}{\overset{R'\quad R}{C}} \quad + Hal^{-} \qquad (211)$$

$$IBr + \quad \underset{R'\quad R}{\overset{R'\quad R}{C}} \quad + Br^{-} \qquad (212)$$

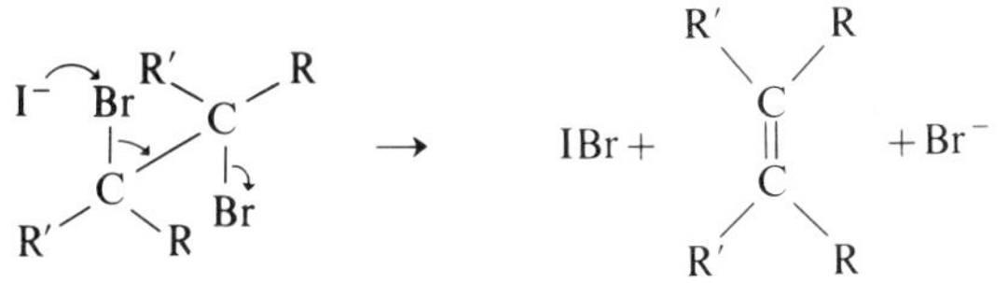

On the basis of the stereochemistry of reagents and products, the two departing atoms are eliminated from positions *anti* to each other, as shown. The absence of significant amounts of *cis* elimination products implies that considerable double bond character is present in the central bond in the transition state.

Cis addition and elimination

The *trans* elimination and addition reactions discussed in the last section have involved radical or ionic intermediates. It is established that bonding can be preserved through the reaction and steric hindrance minimized by reaction in this sense (i.e. *trans*). However, a few additions to double bonds and the corresponding eliminations are *cis*. Notable examples are the addition of BH_3 to olefins (hydroboration) and the gas-phase elimination of hydrogen halides from alkyl halides.

The elimination of hydrogen halide from an alkyl halide by base involves a *trans* loss of hydrogen halide, rationalized in terms of the ionic mechanism discussed earlier. In the gas phase, alkyl halides undergo a first-order loss of hydrogen halide, which appears to be a simple unimolecular reaction. The elimination was shown in 1952 to be *cis* in nature by Barton, Head, and Williams who heated (−)-menthyl chloride **181** to give (+)-menth-2-ene **182** and (+)-menth-3-ene **183** as products along with hydrogen chloride.

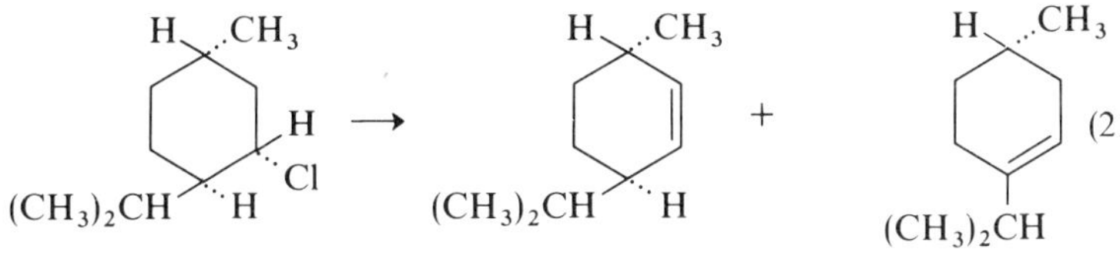

181   **182**   25%   **183**   75%

The production of **183** in 75 per cent yield implies that the elimination is at least 75 per cent *cis*, since if the elimination were *trans* **182** only should be produced. Since neither hydrogen atoms nor chlorine atoms are formed during the reaction, it must be concluded that the reaction is a molecular four-centre process in which the new H—Cl and C=C π bonds are being formed and the C—H and C—Cl bonds break.

Hydroboration, the reaction of alkenes with diborane to give di- or tri-alkylboranes, is a synthetically useful reaction. The addition seems to have electrophilic character, but it is stereospecifically *cis*, suggesting that a molecular reaction is taking place.

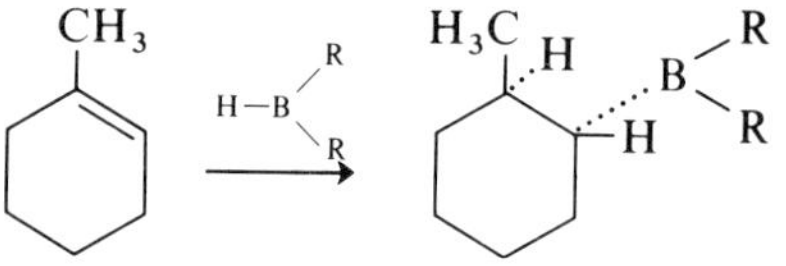

(214)

Concerted molecular reactions

A number of addition, elimination, and rearrangement reactions appear to be purely molecular in nature, with no evidence for ionic intermediates.

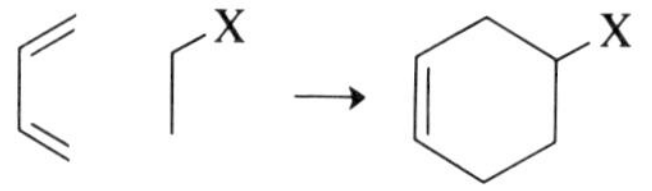

(215)

Cis addition, readily accomplished. Useful preparatively.

(216)

Does not go at all. The activation energy for the reaction would probably be at least 182 kJ/mol^{-1}.

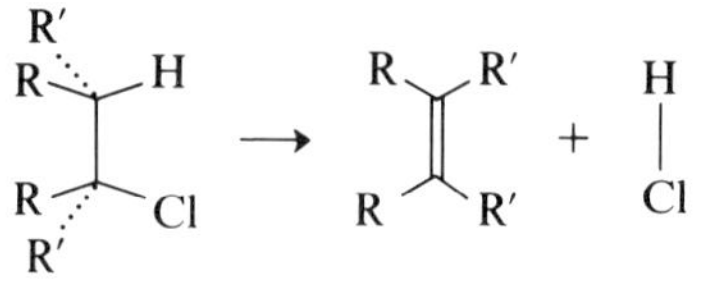

(217)

Cis. Very high activation energy required.

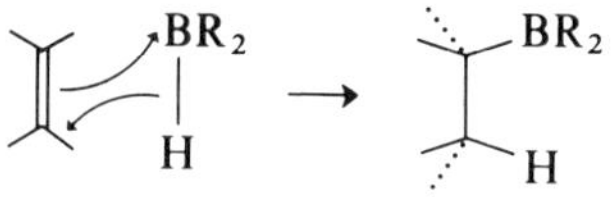

(218)

Cis. Occurs at room temperature or on slight warming.

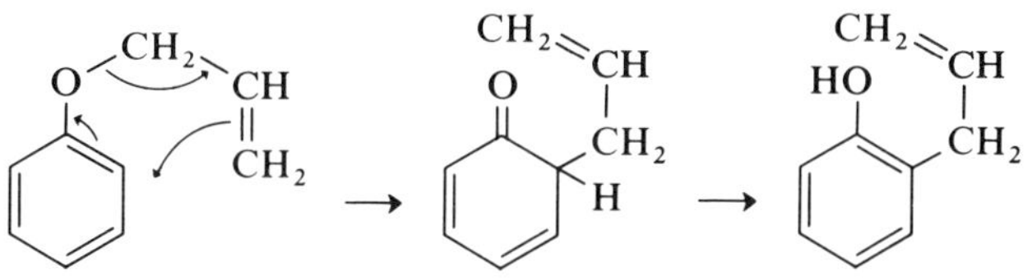

(219)

Occurs on heating. Fairly facile in view of the fact that the aromatic stabilization energy is lost in the intermediate.

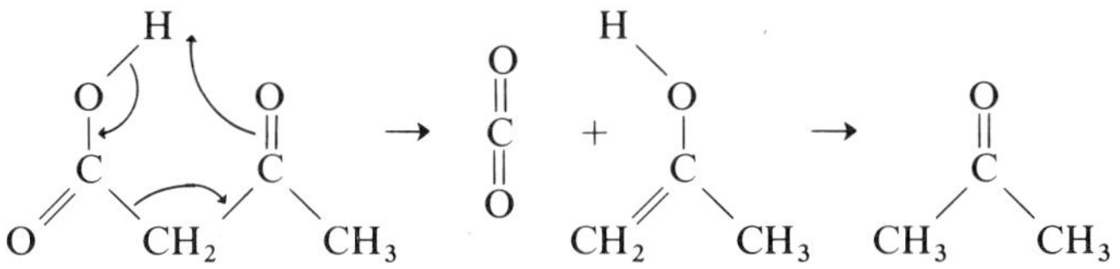

(220)

Occurs readily on heating.

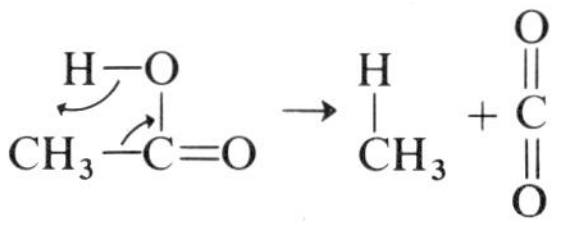

(221)

Occurs only above 500°.

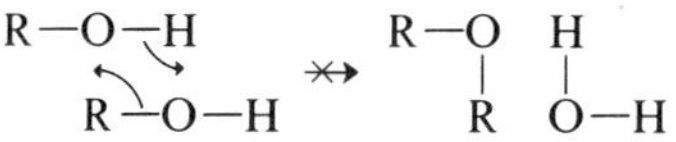

(222)

Almost without exception organic functional group reactions do not go this way.

Addition reactions in this category are usually *cis* in nature; hence it seems reasonable to assume that both new bonds are formed at the same time. An important example is the Diels–Alder reaction in which a diene reacts with an (activated) olefin to give a cyclohexene derivative (reaction 215). A number of examples of molecular reactions are shown in the chart. It will be seen that some of these reactions occur very readily whereas others are very slow or do not take place. A simple view would be that in these reactions bonding is preserved during the reaction by synchronous breaking and formation of new bonds. Since equal numbers of bonds are being made and broken this should lead to transition states which are not too high in energy. However, this simple picture is not entirely adequate.

The most obvious general feature of these reactions is that concerted reactions involving three bonds broken and formed seem to be favourable, whereas those involving only two bonds range from the readily performed to the impossible. However in considering reactions of this type it is important to consider not the individual bonds or orbitals, but *all* the orbitals involved, since in the transition state molecular orbitals of the entire reacting system will be important. For the systems involving three bonds broken and formed, i.e. six orbitals, the transition state will involve partial overlap of all these orbitals with their nearest neighbours in a six-membered ring. This will give rise to a benzene-like situation (six overlapping orbitals, six electrons) where there will be three low-lying molecular orbitals which can accommodate the six electrons. Hence the transition state will be stabilized by this benzene-like resonance and will therefore be relatively accessible energetically from the reactants.

A similar planar transition state for the four-electron systems would give rise to a cyclobutadiene-like transition state. Since cyclobutadiene is not resonance-stabilized these transition states will be much higher in energy than the corresponding reactant molecules and these reactions will only occur with difficulty if at all. Thus reactions of types (216, 217, 221, 222) will be difficult or impossible.

Reaction (223) is interesting from this point of view. By the Hückel rule, cyclic π-systems with $4n+2$ electrons are aromatic, whereas systems with $4n$ electrons are anti-aromatic. If however a π-system could be constructed where there was negative overlap between one pair† of adjacent orbitals and positive overlap between all the others, then the situation would be reversed: $4n$ electrons would now be the preferred number. It is difficult to construct carbocyclic compounds of this type (sometimes called Möbius systems), but more complex systems can show this type of behaviour and transition states of this type can often be achieved. In reaction (223), the cyclobutene ring can open in two ways: the two orbitals which comprised the breaking sigma bond may either rotate in the same direction (conrotatory ring opening) or in the opposite directions (disrotatory). The mode of ring opening can be deduced from the products. If disrotatory ring opening takes place, a Hückel type transition state (**185**) is involved, whereas the conrotatory mode gives a Möbius type transition state **184** with one pair of orbitals

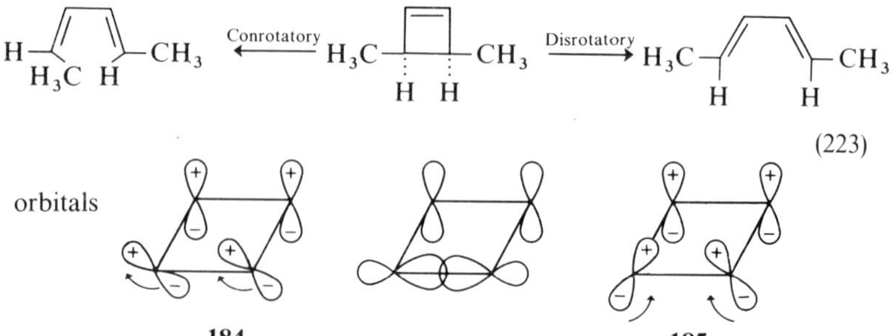

(223)

orbitals

184   **185**

† Or an odd number of pairs.

having negative overlap. Since there are four electrons involved in the system, the Möbius transition state will have lower energy, and conrotatory ring opening occurs.

Hydroboration

Hydroboration, reaction (218), at first sight appears anomalous in that it seems to be a four-centre reaction of the Hückel type, but with a low activation energy. However, boron compounds are unusual (in relation to most organic compounds) in having an empty valence-shell orbital. The presence of this extra empty orbital makes it possible that hydroboration may be concerted (i.e. both new bonds are made at the same time) but that the reaction does not truly involve a cyclic transition state, since the new bond to boron may use a different boron orbital from the one used in the breaking B—H bond (**186**). Thus a concerted reaction can take place

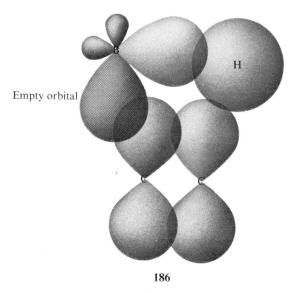

Empty orbital

186

without going through a cyclic high-energy cyclobutadiene-like intermediate. Molecules like HCl do not have a vacant valence-shell orbital. Hence the same orbitals must be used in the reactants as in the products and thus the molecular addition would have to go via a high-energy cyclobutadiene-like transition state.

Biradical intermediates

The activation energy required for some of the 'forbidden' four-centre reactions approaches or even exceeds the energy required to break one of the covalent bonds in the system. If this is so, an alternative reaction path via a biradical intermediate may be possible. There is evidence that the ring opening of cyclobutane to two molecules of ethylene may occur via the biradical intermediate **187**, rather than as a direct four-centre reaction (225). Evidence for a biradical intermediate in a reaction of this type carried out in the

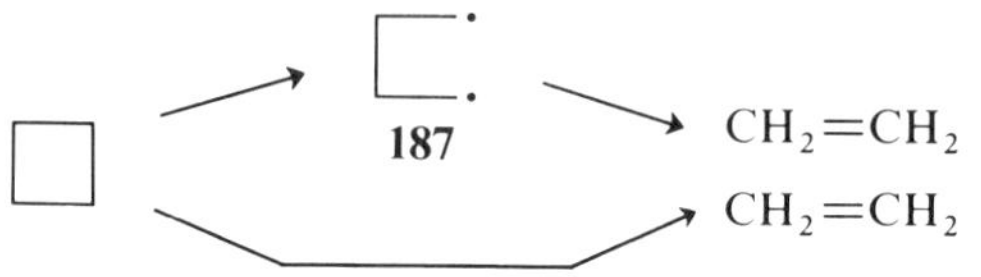

reverse direction was obtained by Bartlett's group in 1964. They heated 1,1-dichloro-2,2-difluoroethylene **188** with *trans,trans*- and *cis,trans*-hexa-2,4-diene (**189** and **190**). In this system the intermediate radical is stabilized by the conjugated double bond, which makes the radical allylic. It appears that rotation of the intermediate radical competes with ring closure, since although both **189** and **190** give more **193** than **194**, the **193/194** ratio is higher (5·1 : 1) when the reaction is carried out with **189** than with **190** (3·2 : 1).

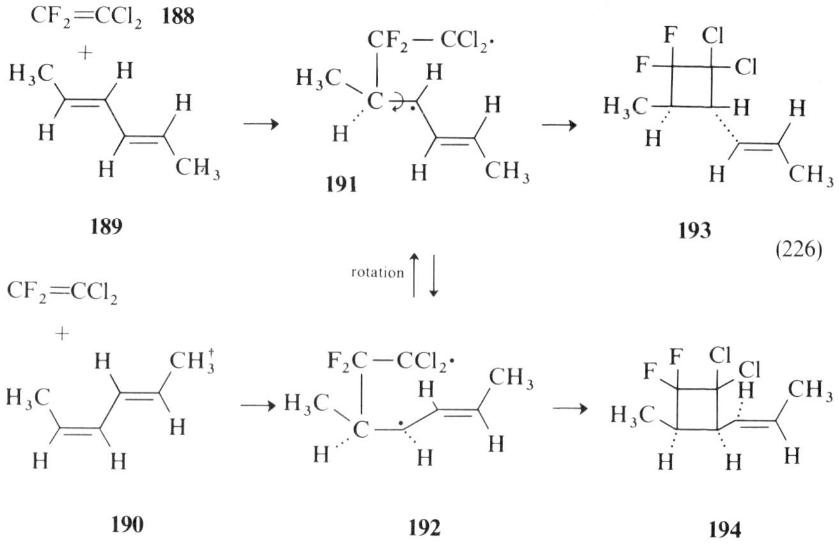

† Attack on the other double bond of this molecule also takes place.

Interpretation of data on new addition and elimination reactions

Reactions found to give *trans* addition to a double bond will usually have involved an intermediate (probably radical or cationic). If a *cis* addition has taken place, a molecular reaction should be suspected.

If the proposed mechanism involves a cyclic transition state, the mechanism should be checked to see if its transition state is 'aromatic' or not. If not, then there should be a high activation energy (of the order of magnitude of the weakest bond in the system).

PROBLEMS

5.1. $(+)$-Ph Np MeSi·O·CO·CH$_3$ is converted by potassium hydroxide in xylene into $(-)$-Ph Np MeSi·OK, which in turn may be converted by acetyl chloride into $(-)$-Ph Np MeSi·O·CO·CH$_3$. What can be said about the stereochemistry of these two reactions?

5.2. S_N2 reactions of *secondary* alkyl halides take place with inversion of configuration. How could you show that inversion also takes place during S_N2 reactions of *primary* halides?

5.3. Pure trialkylboranes rearrange on heating to give a mixture of isomers in which the compounds with boron attached to primary carbon atoms predominate. For example $(3\text{-}C_6H_{13})_3B \rightarrow (1\text{-}C_6H_{13})_3B$. Suggest a mechanism which accounts for the fact that the boron atom can move along the carbon chain but cannot pass a quaternary carbon atom, e.g. the grouping $C—CMe_2—C$.

5.4. Explain why menthyl chloride **181** should give exclusively 2-menthene **182** when treated with alcoholic KOH.

5.5. On heating at 260°C for five hours, $(+)$-Ph Me EtC·NC rearranges to Ph Me EtC·CN which is optically inactive. Suggest a plausible mechanism, and also a desirable control experiment.

6. Other evidence

CHAPTERS 2 to 5 describe the main methods used in studying reaction mechanisms. There are a number of other techniques which are less frequently used in deducing reaction mechanisms but which often provide further evidence that a particular mechanism is reasonable (or not) and give a more detailed insight into the reaction mechanism.

Steric acceleration and deceleration

Many reactions involve a change in the number of groups coordinated to a particular atom on going to an intermediate or transition state. For example in an S_N2 reaction the original tetravalent carbon acquires an extra group in the transition state, whereas in an S_N1 reaction the intermediate is a carbonium ion with only three attached groups. Hence, other things being equal, if the substituents R are bulky (i.e. not hydrogen), S_N1 reactions should be favoured.

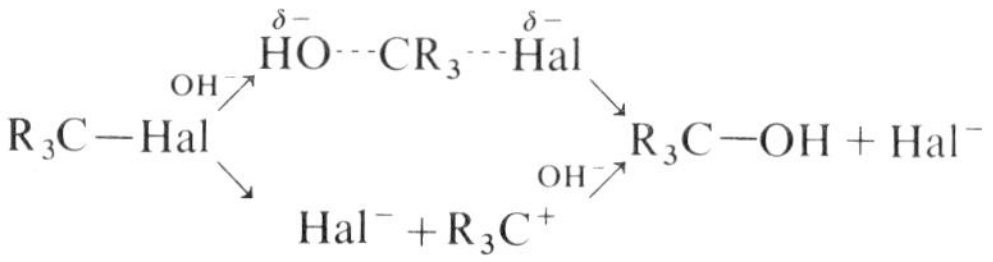

$$R_3C-Hal \xrightarrow[\text{OH}]{\overset{\delta-}{\text{HO}}\cdots CR_3\cdots \overset{\delta-}{\text{Hal}}} R_3C-OH + Hal^- \qquad (227)$$

$$Hal^- + R_3C^+$$

In accordance with this, methyl and ethyl bromides undergo S_N2 reactions almost exclusively. Isopropyl bromide hydrolyses by both paths, and t-butyl bromide reacts mainly by the S_N1 route.

The acid-catalysed hydrolysis of esters usually goes via the following route.

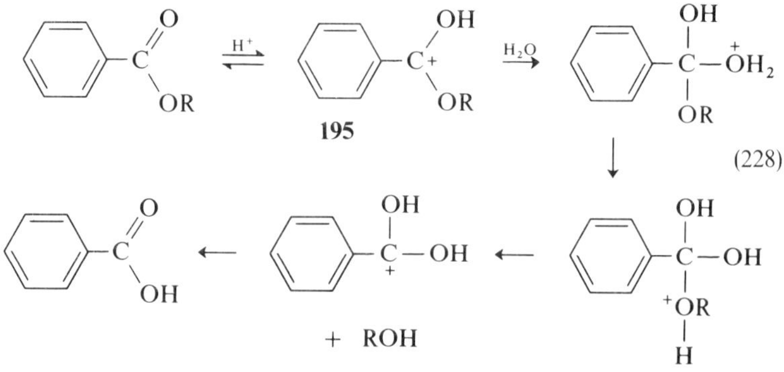

$$(228)$$

When a water molecule attacks the carbonium-ion intermediate **195**, the coordination around the central carbon atom is increased from three to four. In concentrated sulphuric acid, the more crowded ester **196** is hydrolysed by the alternative route (229), in which

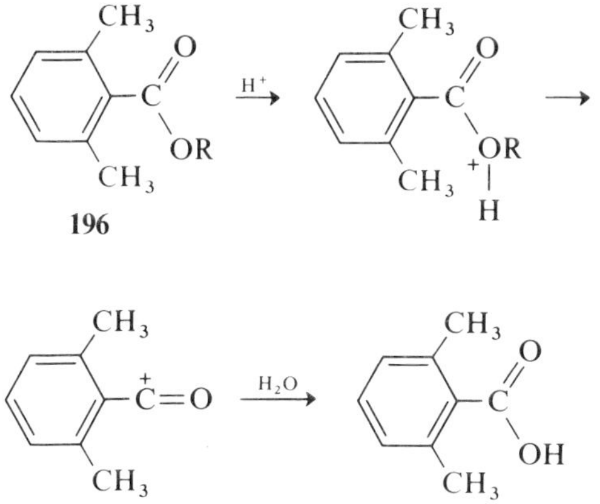

196

(229)

197

steric crowding is reduced when the intermediate acylium ion, **197**, with only two-coordinated carbon is formed.

Steric effects can rarely be isolated from other possible effects of substituents (see next section).

Substituent effects

Transition states or intermediates often develop charge or radical character. If this happens, appropriate substituents at the reaction centre may stabilize or destabilize the transition state or intermediate.

Stabilization of cations

Experimentally it is found that carbonium-ion formation is facilitated by substitution of hydrogen by alkyl groups. Thus if n-propyl fluoride is dissolved in antimony pentafluoride, the originally formed n-propyl cation rearranges to the more stable isopropyl cation. Tertiary cations are more stable still: t-butyl bromide dissolves in liquid SO_2 to give conducting solutions which contain the t-butyl cation: primary and secondary halides ionize less readily. Groups such as CF_3 destabilize carbonium ions. The stabilizing effect alkyl > H > CF_3 is rationalized partly in terms of an *inductive* effect: some groups (CH_3) can more effectively release their σ electrons to stabilize a cation than can others (CF_3) where there is polarization of electrons *towards* the CF_3 group which tends to destabilize the carbonium ion.

This stabilization is an experimental fact—it does not matter for many purposes how this stabilization can be partitioned among the various possible causes (inductive, steric, hyperconjugative effects, bond strength changes with hybridization etc.). If alkyl groups at a reaction centre make a reaction faster, whilst CF_3 groups make it slower, this is good evidence for an intermediate with some carbonium-ion character.

For example, the addition of hydrogen halides to olefins, when carried out with purified reagents, is thought to involve carbonium ion intermediates such as **198**, **199**, and **200**. It is found that propene reacts faster than ethylene with hydrogen halides, and gives the isopropyl compound (reaction 231) rather than the n-propyl compound. 3,3,3-Trifluoropropene on the other hand reacts more slowly than ethylene, and gives the primary product. These relative rates and orientations accord with the formation of a carbonium-ion intermediate. In reaction (231), the iso-propyl cation **199** is more stable than the n-propyl cation, which explains the preferred direction of addition, and the iso-propyl cation **199** is also more stabilized than the ethyl cation **198**, which explains why addition to propene is faster than to ethylene. In reaction (232) the presence of the CF_3 group destabilizes either of the possible cations **200** or **201**; hence the reaction is slower than (231). However the primary cation **200** has the destabilizing CF_3 group further away than the alternative cation **201**; thus the primary bromide **202** is formed as the product.

$$CH_2{=}CH_2 \xrightarrow{HBr} CH_3\overset{+}{C}H_2 \rightarrow C_2H_5Br \qquad (230)$$
$$\mathbf{198}$$

$$CH_3{-}CH{=}CH_2 \xrightarrow{HBr} CH_3{-}\overset{+}{C}H{-}CH_3 \rightarrow CH_3{-}CHBr{-}CH_3 \quad (231)$$
$$\mathbf{199}$$

$$CH_3{-}CH_2{-}\overset{+}{C}H_2$$
less stable

$$CF_3{-}CH{=}CH_2 \xrightarrow{HBr} CF_3{-}CH_2{-}\overset{+}{C}H_2 \rightarrow CF_3{-}CH_2{-}CH_2Br \quad (232)$$
$$\mathbf{200} \qquad\qquad \mathbf{202}$$

$$CF_3{-}\overset{+}{C}H{-}CH_3$$
$$\mathbf{201}$$
less stable

Free radicals

Alkyl substituents stabilize radical centres, but the effects are less pronounced than at carbonium-ion centres since no formal positive charge has to be stabilized. It is found experimentally that tertiary radicals are more stable than secondary, which in turn are more stable than primary. For example, the bond dissociation energies $D(R—H)$ increase in the order $D(Bu^t—H) = 380$; $D(Pr^i—H) = 395$; $D(Et—H) = 410$; $D(Me—H) = 437$ kJ mol^{-1}. Since there is good evidence that for example

$$D(Et—H) \sim D(Pr^n—H)$$

it is clear that secondary radicals are more stabilized than primary radicals, whereas tertiary radicals are more stable still. These differences in radical stability provide, among other things, further understanding of the free-radical chain reaction of hydrogen bromide with olefins where it is found that propene reacts faster than ethylene and gives 1-bromopropane as the product (reaction 233). Addition of a bromine atom to propene gives the radical **203**, rather than the less stable primary radical **204**, thereby determining the direction of the addition; the radical **203** is also more stabilized than the radical $\cdot CH_2CH_2Br$, so that addition to propene is faster than to ethylene.

$$CH_3—CH{=}CH_2 + Br\cdot \rightarrow CH_3—\overset{\cdot}{C}H—CH_2Br \overset{HBr}{\rightarrow} CH_3CH_2CH_2Br + Br\cdot$$

203   (233)

$$\rightarrow CH_3—CHBr—CH_2\cdot$$

204

Conjugation

Substitution of unsaturated groupings such as $C{=}C$, $C{=}O$, or Ph at a reaction centre should facilitate reactions involving carbonium ions, carbanions, or free radicals (it also speeds up S_N2 reactions for example). In view of this lack of selectivity it is perhaps not a very good diagnostic test for a particular type of reaction. However, it may have a negative predictive effect: if a reaction is postulated to go via a carbonium-ion intermediate, and the reaction does not go faster with a phenyl substituent at the supposed site of the carbonium ion, then the mechanism will be suspect. The effect of the conjugating group is to spread the charge or free electron density: the delocalized structure is more stable than the (unsubstituted) localized structure.

$$\langle\!\!\!\!\bigcirc\!\!\!\!\rangle—\overset{+}{C}H_2 \leftrightarrow +\langle\!\!\!\!\bigcirc\!\!\!\!\rangle{=}CH_2 \text{ etc.} \qquad (234)$$

Electronic effects of substituents: the Hammett equation

A number of attempts have been made to put substituent effects onto a quantitative basis. One of the most successful of these schemes derives from the acidities of a series of m- and p-substituted benzoic acids which can be measured very accurately. A substituent constant σ was defined for each substituent by equation (236).

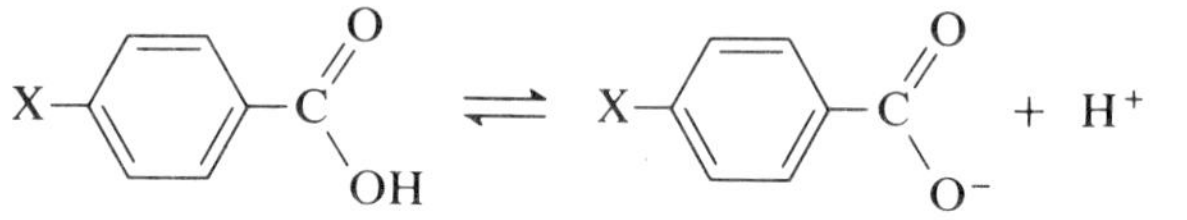

205

$$\log \left[\frac{K_a(X \cdot C_6H_4 \cdot CO_2H)}{K_a(C_6H_5 \cdot CO_2H)} \right] = pK_a(C_6H_5 \cdot CO_2H) - pK_a(X \cdot C_6H_4 \cdot CO_2H) = \sigma$$
$$(236)$$

Since electron-withdrawing substituents will stabilize the anion of the acid, they will increase its acid strength, and be associated with positive values of σ. Conversely, electron-releasing substituents will have negative values of σ. A list of some σ constants is given in Table 5.

How can these substituent constants be used for other reactions of aromatic compounds? It was found that, for a very wide variety of reactions, if rates (or equilibrium constants) were measured for a number of substituted

TABLE 5

Some values of Hammett substituent constants[†]

Substituent	meta	para		
	σ	σ	σ^+	σ^-
NMe_2	-0.21	-0.83	-1.70	
CH_3	-0.07	-0.17	-0.31	
CH_3O	0.11	-0.27	-0.78	
F	0.34	0.06	-0.07	
Cl	0.37	0.23	0.11	
Br	0.39	0.23	0.15	
I	0.35	0.18	0.13	
CO_2Et	0.37	0.45	0.48	0.68
CF_3	0.42	0.54	0.61	
CN	0.56	0.66	0.66	0.87
NO_2	0.71	0.78	0.79	1.24

[†] For further values see for example C. D. Ritchie and W. F. Sager, *Prog. Phys. Org. Chem.*, 1964, **2**, 323.

compounds ArX, as well as for the unsubstituted compound ArH, then a graph of $\log(k_{ArX}/k_{ArH})$ against σ gave a straight line, which implies that a relationship of type (237) holds.

$$\log(k_{ArX}/k_{ArH}) = \rho\sigma \qquad (237)$$

This equation is known as the Hammett equation. The slope of the graph, ρ, is characteristic of the particular reaction and is a measure of the demand of the transition state for electrons. Benzoic acids are made stronger by electron-withdrawing substituents and since ρ for the acid dissociations of benzoic acids is unity by definition, positive values of ρ for a reaction imply a transition state where negative charge is being accumulated or an initial positive charge is dispersed. Conversely, reactions where positive charge is being accumulated at the reaction centre in the transition state should have negative values of ρ. A few examples of reactions with their ρ constants are listed in Table 6.

TABLE 6

Hammett ρ values

Reaction	Medium	
(235) $ArCO_2H \rightleftharpoons ArCO_2^- + H^+$	water	1·0(
(245) $ArCO_2H \rightleftharpoons ArCO_2^- + H^+$	ethanol	1·9(
(246) $ArCH_2CO_2H \rightleftharpoons ArCH_2CO_2^- + H^+$	water	0·49
(247) $ArNH_3^+ \rightleftharpoons ArNH_2 + H^+$	water	2·77
(248) $Ar_3CCl \rightleftharpoons Ar_3C^+ + Cl^-$	SO_2	$-3·97$
(249) $ArCMe_2Cl + H_2O \rightarrow ArCMe_2OH$ $+ HCl$	acetone/water	$-4·54$
(250) $ArH + HNO_3 \rightarrow ArNO_2 + H_2O$	acetic anhydride	$-7·29$
(251) $ArH + HOBr \rightarrow ArBr + H_2O$	perchloric acid/dioxan/water	$-6·2$
(252) $ArH + Br_2 \rightarrow ArBr + HBr$	acetic acid/water	$-12·1$
(253) $ArCl† + MeO^- \rightarrow ArOMe + Cl^-$	methanol	$+3·9$
(254) $ArO^- + CH_2 \overset{O}{-} CH_2$ $\rightarrow ArOCH_2CH_2O^-$	ethanol	$-1·12$
(255) $ArCH_2\cdot \rightarrow ArCH_2^+ + e^-$	gas	-20
(256) $ArCH(Cl)CH_3 \rightarrow ArCH{=}CH_2 + HCl$	gas	$-1·36$
(257) $ArCH_3 + Cl\cdot \rightarrow ArCH_2\cdot + HCl$	carbon tetrachloride	$-0·66$
(258) $ArCO\cdot Cl + Bu_3^nSn\cdot \rightarrow ArCO\cdot + Bu_3^nSnCl$	m-xylene	$+2·6$

$†\ ArCl = X\!-\!\!\left\langle\!\!\bigcirc\!\!\right\rangle\!\!-\!Cl$ with NO_2

σ^+ *and* σ^-

The stabilization or destabilization of the substituted benzoate ion **205**, which is responsible for the variation in acidities of benzoic acids, is mainly by the inductive effect (polarization of electrons along σ bonds), since the negative charge in the benzoate ion is not conjugated with the aromatic ring. (Reasonable resonance structures with the negative charge on the ring cannot be drawn). However, some reactions produce charges which are conjugated with the aromatic group, and for these reactions, delocalization on to the ring and on to certain *para* substituents will be important. Thus groups such as MeO and NMe$_2$ are effective in stabilizing positive charges, whereas groups such as CO$_2$Me, CN and NO$_2$ are effective in stabilizing negative charges. Typical resonance structures are shown below for substituted benzyl cations **206** and phenoxy anions **207**. Such resonance effects are not shown by substituents at the *meta* positions.

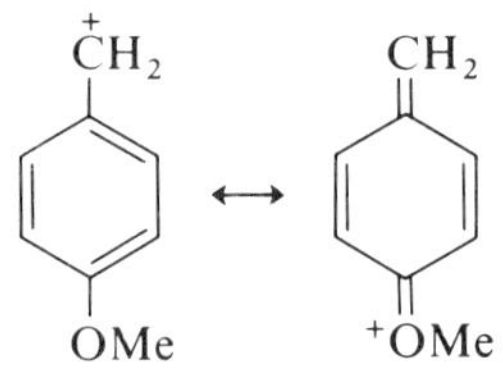

206 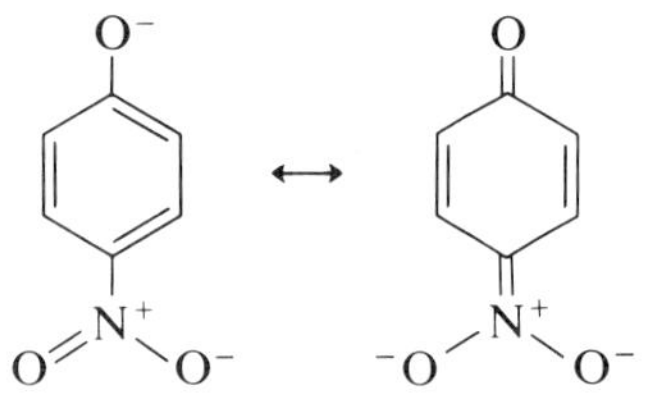   **207**

Two further standard reactions have been suggested to provide substituent constants appropriate to situations where resonance effects are important. These are the acid dissociations of substituted phenols (equilibrium 238), and the rates of solvolysis of substituted cumyl chlorides in aqueous acetone (reaction 239). The ρ values for these two reactions were established by consideration of *meta* substituted compounds only, and the ρ values were combined with the rates (or equilibrium constants) for the *para* substituted compounds to define a series of σ^- and σ^+ values, which should be more appropriate than σ values for application to other reaction systems where positive or negative charge is conjugated with the aromatic ring during a reaction.

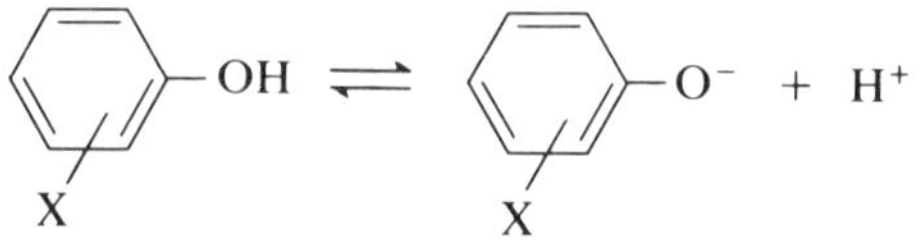

(238)

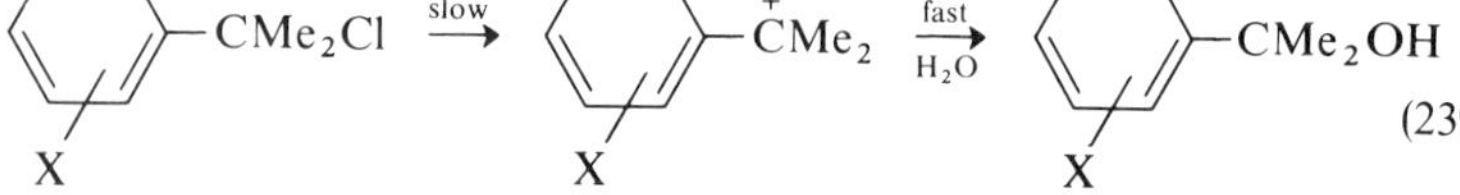

(239)

Mechanistic use of the Hammett equation

The Hammett equation may be used to give information about the mechanism of a reaction. The sign and magnitude of ρ are of interest, and also which of the three substituent constants σ, σ^+, or σ^- gives the best Hammett plot. This gives information about the transition state of the reaction, summarized in Table 7. Occasionally the category of the reaction in Table 7 allows a choice between two possible mechanisms to be made: more usually the category confirms an already postulated mechanism, and the magnitude of ρ provides more quantitative information about the charge developed in the transition state.

As an example of the use of the Hammett equation in deciding between alternative reaction mechanisms, two possible mechanisms for the acid-catalysed rearrangement of N-nitro-N-methylanilines to the corresponding o-nitro-N-methylanilines are shown in schemes (240) and (241).

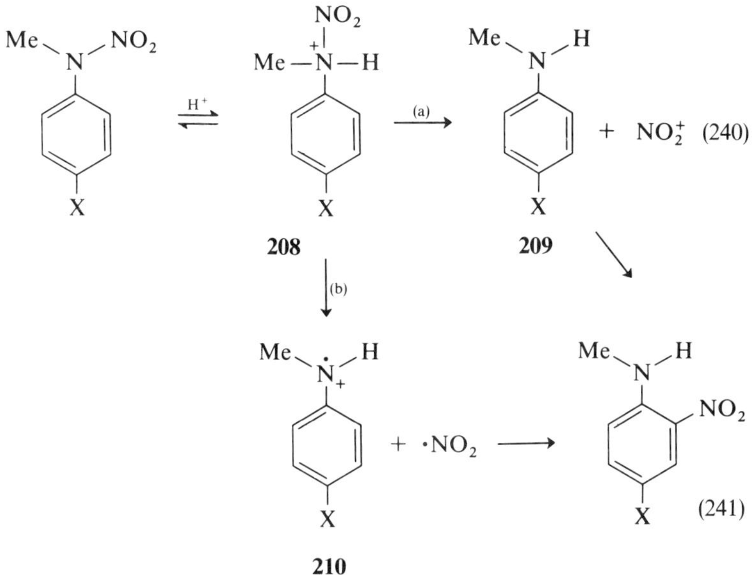

If reaction (a) is rate-determining, a positive ρ value would be expected, and a correlation with σ, since **208** should be stabilized (inductively) by electron-releasing substituents to a greater extent than the uncharged methylaniline **209**. On the other hand, if route (b) is involved, conjugation of the positive charge with the ring in **210** will give greater stabilization than the inductive stabilization of the reactant **208**. Hence a negative ρ value would be expected, and a correlation with σ^+ should be observed for this route. In fact, a ρ value

TABLE 7

Deductions made from the sign of ρ, and which of σ, σ^+ and σ^- gives the best fit in the Hammett equation†

	ρ positive	ρ negative
σ^-	Reaction centre§ conjugated with Ar group—negative charge built up at reaction centre in transition state. Example: reaction 265.	Reagent stabilized by conjugative release of electrons to Ar group—this conjugation less important in transition state. Example: reaction 266
σ	Reaction centre not conjugated with Ar group—transition state has less demand than reactant for electrons. Example: hydrolysis of aromatic esters by bases	Reaction centre not conjugated with Ar group—transition state has greater demand than reactant for electrons. Example: $Ar_2C{=}\overset{+}{N}{=}\overset{-}{N} + PhCO_2H$ $\rightarrow Ar_2CH{\cdot}CO_2Ph + N_2$
σ^+	Reaction centre conjugated with Ar group and stabilized by electron-releasing substituents—conjugation less important in transition state. Examples: rare	Reaction centre conjugated with Ar group—transition state has greater demand than reactant for electrons. Example: reaction 264

† σ values are used where a special σ^+ or σ^- value has not been derived.

§ The reaction centre for the purpose of this table is the position at which charge is built up or reduced during the reaction.

of $-3{\cdot}7$, with correlation best against σ^+, was found experimentally, providing strong support for route (241) rather than (240).

Hammett plots sometimes cast light on the nature of an intermediate in a reaction. Free-radical bromination of toluenes by molecular bromine takes place by a chain reaction (242). Similar brominations can be carried out by using *N*-bromosuccinimide **211** instead of bromine, and the mechanism was originally thought to be (243). More recently, it has been suggested that the function of the *N*-bromosuccinimide is to provide a small quantity of molecular bromine, and to maintain it by reaction (244) (which is known to be fast) while the actual bromination is carried out by the molecular bromine (reaction 242). This mechanism is supported by ρ values obtained for substituted toluenes when compounds **211**, **212**, **213**, and molecular bromine were used as the brominating reagents. The ρ values are identical within experimental error even though **213** has four fluorine substituents, whereas **212** has four methyl substituents, and bromine is an entirely different compound. It is most unlikely that the radicals derived from all four brominating agents would attack substituted toluenes with the same ease: the only likely explanation is that only one intermediate, the bromine atom, is involved, and that the mechanism of bromination involves (244) and (242), not (243).

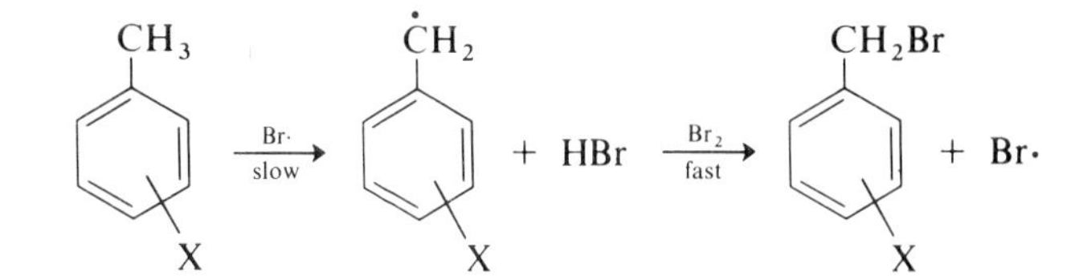

$$\text{(242)}$$

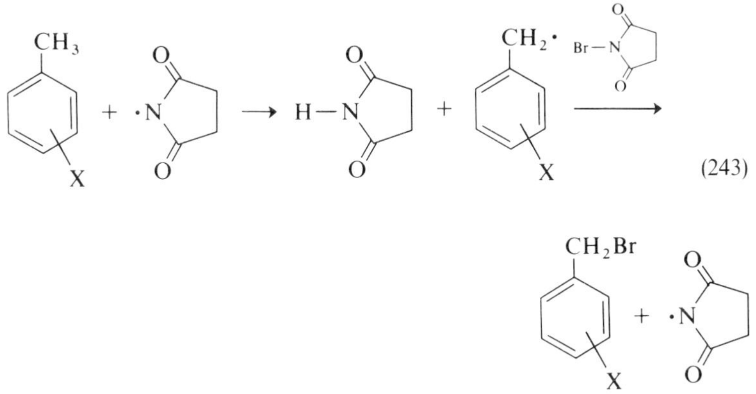

$$\text{(243)}$$

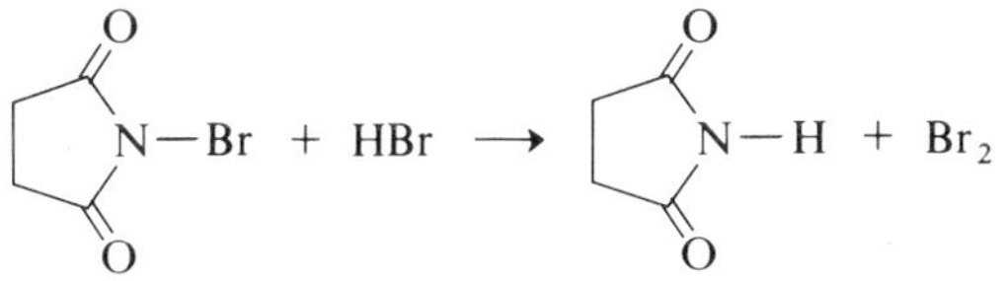

$$\text{(244)}$$

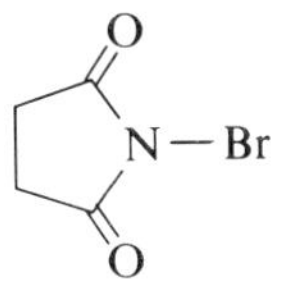

211	**212**	**213**	
$\rho = 1\cdot46\pm0\cdot07$	$1\cdot45\pm0\cdot07$	$1\cdot36\pm0\cdot05$	$1\cdot36\pm0\cdot5$

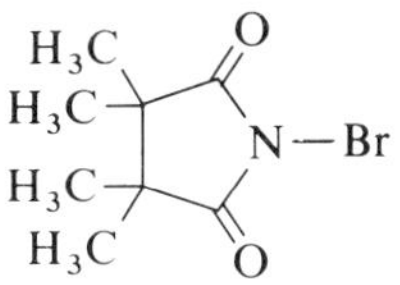

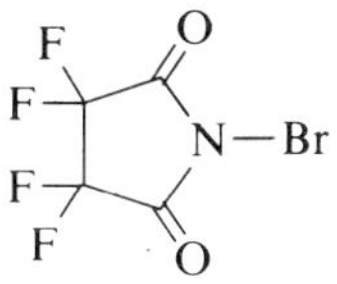

The magnitude of the ρ factor is a measure of the amount of charge accumulated at the reaction centre in the transition state. The highest observed values are for ionizations carried out in the gas phase in a mass spectrometer (reaction 255), where there is no solvent to stabilize the ions, and therefore the stabilizing effect of substituents is at a maximum. Smaller ρ factors will be found

(a) in solution in polar solvents—ρ for benzoic acid dissociations is higher in ethanol than in water;

(b) if the aromatic ring is not adjacent to the developing charge—ρ for phenylacetic acid is approximately half the value for benzoic acid;

(c) if the transition state only involves a small extent of build up of charge, as in the radical reactions (257) and (258)—Table 6, p. 107.

When assessing the amount of charge developed during a reaction, it is helpful to consider the ρ value obtained in comparison with ρ values obtained for reactions *in a similar solvent* which involve a complete separation or destruction of charge, and which therefore should give maximum ρ values for the solvent considered. Examples of such reactions are the solvolysis of cumyl chloride (reaction 249) and the ionization of benzyl radicals (reaction 255). Thus in assessing the charge separation in the transition state of reaction (256), the ρ value of $-1\cdot36$ at 335°C, which would correspond to a value of $-2\cdot6$ at 25°C, should be compared with the value of -20 obtained for

$$Ar-\underset{\underset{Cl}{|}}{\overset{}{C}}H-\underset{\underset{H}{|}}{\overset{}{C}}H_2 \rightarrow Ar-\underset{\underset{\underset{\delta-}{Cl}}{\overset{\delta+}{|}}}{\overset{}{C}}H\text{=}\underset{\underset{H}{\cdots}}{\overset{}{C}}H_2 \rightarrow Ar-CH\text{=}CH_2 \quad (256)$$
$$Cl-H$$

reaction (255). This shows that charge separation in (256), though significant, does not approach complete ionization.

Solvent effects

Organic acids such as benzoic acid are more highly ionized in aqueous or alcoholic solution than in say acetone or hexane solution. Water and alcohol stabilize anions and cations by solvation, and thus the dissociation of a neutral organic acid into ions is favoured in these solvents, which we may term ionizing solvents as opposed to non-ionizing solvents such as acetone and hexane. (It may be noted that good ionizing solvents have high dipole moments, and are also hydrogen bonded).

$$HA \quad\quad \rightleftharpoons \quad\quad H^+ \quad A^- \quad\quad (259)$$

nature of solvent has less effect on undissociated acid than on ions	stabilized by ionizing solvents

Hughes and Ingold argued that the same type of effect would operate in chemical reactions, and that in general, if charge is built up in the transition state relative to the starting materials, then the reaction rate will be substantially increased by carrying out the reaction in a more highly ionizing solvent. Conversely, if charge is destroyed in the transition state the reaction rate will be decreased if the solvent is made more highly ionizing. In some reactions charge is not created or destroyed, but is spread in the transition state. For example in nucleophilic substitutions such as the second reaction in Table 8 the charge is spread over the incoming and outgoing groups. Such reactions will be favoured by non-ionizing solvents, since in an ionizing solvent the initial ionic reagents will be more stabilized than the transition state, but the effect will be smaller than for reactions where charge is created or destroyed.

This principle can be employed to cover a wide range of reactions involving species of different charge. Examples are shown in Table 8 (p. 114).

Changing the solvent thus provides a useful check on a postulated mechanism. To be applied effectively, it is best to alter the ionizing power of the solvent only slightly, for example by changing from 80:20 to 70:30 acetone/water, since then there is little danger of changing the mechanism. A more drastic change of solvent (e.g. from acetone to water) might have the effect of changing the mechanism of the reaction, for example a change of mechanism from S_N2 to S_N1 might take place.

Hammond's postulate

We conclude this chapter with a concept which provides insight into variations of reactivity, rather than gross differences in mechanism. We may represent exothermic, thermoneutral, and endothermic reactions qualitatively as in Fig. 6, where the energy of the reaction system has been plotted against an arbitrary reaction coordinate.

Hammond's postulate states that for reactions which are very exothermic, the transition state will closely resemble the starting materials (Fig. 6a), whereas for reactions which are very endothermic, the transition state will resemble the products (Fig. 6c). For reactions which are approximately thermoneutral, the transition state will be a kind of 'half-way house' between starting materials and products, with the bonds to be broken and those to be formed all at about half strength. This concept can be usefully applied to a wide variety of organic reactions. For reactions involving intermediates, the concept applies to the individual reactions, not to the overall process.

For reactions which are approximately thermoneutral, this postulate implies that any particular stabilization or destabilization of a product relative to the starting material, will be present to approximately half that extent in the transition state. As an example, let us consider the hydrogen-transfer reactions of methyl radicals with methane and with ethane. The thermoneutral reaction of the methyl radical with methane has an

	ΔH	E	
$CH_3 \cdot + H{-}CH_3 \rightarrow CH_3{-}H + CH_3 \cdot$	0	61	(260)
$CH_3 \cdot + H{-}CH_2CH_3 \rightarrow CH_3{-}H + \cdot CH_2CH_3$	-28	49	(261)

TABLE 8

Solvent effects on reaction rates

Reagents	Transition state	Products	Type	Effect on rate of increasing solvent polarity
$(CH_3)_3C-Br \rightarrow (CH_3)_3C^+Br^-$	$(CH_3)_3C^+Br^-$	$\xrightarrow{H_2O} (CH_3)_3C-\overset{+}{O}H_2 + Br^-$	Charge created	Large increase
$HO^- + CH_3-Br \rightarrow$	$\overset{\delta-}{HO}\cdots CH_3\cdots\overset{\delta-}{Br}$	$\rightarrow HO-CH_3 + Br^-$	Charge spread	Small decrease
$H_2O + \overset{+}{S}(C_2H_5)_3 \rightarrow$	$\overset{\delta+}{H_2O}\cdots C_2H_5\cdots\overset{\delta+}{S}(C_2H_5)_2$	$\rightarrow H_2\overset{+}{O}C_2H_5 + S(C_2H_5)_2$	Charge spread	Small decrease
$HO^- + \overset{+}{S}(C_2H_5)_3 \rightarrow$	$\overset{\delta-}{HO}\cdots C_2H_5\cdots\overset{\delta+}{S}(C_2H_5)_2$	$\rightarrow HOC_2H_5 + S(C_2H_5)_2$	Charge destroyed	Large decrease

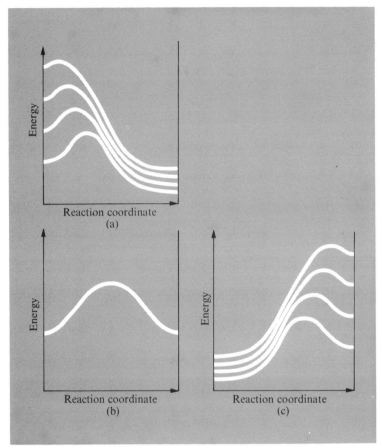

FIG. 6. Hammond's postulate. (a) Exothermic reactions. (b) Thermoneutral reactions. (c) Endothermic reactions.

activation energy of 61 kJ mol^{-1}. The similar reaction of methyl radicals with ethane involves attack at a C—H bond which is approximately 28 kJ mol^{-1} weaker than the C—H bond in methane. In the transition state the C$_2$H$_5$—H bond will be approximately half broken, and hence the *difference* in energies between the two transition states for reactions (260) and (261) should be approximately 14 kJ mol^{-1}: in close agreement with the experimental difference of 12 kJ mol^{-1}.

For endothermic reactions, the transition state will resemble the products more closely, and any stabilizing influences in the products will be more strongly reflected in the transition state. Thus the reactions of bromine atoms with methane and with ethane (262 and 263) are substantially endo-thermic: in this pair of reactions, the activation energy difference is 21 kJ mol^{-1} which is 75 per cent of the difference in enthalpies of the two reactions. In reactions which are extremely endothermic, such as the thermal

$$\begin{array}{lccc}
 & \Delta H & E & \\
\text{Br} \cdot + \text{H} - \text{CH}_3 \rightarrow \text{Br} - \text{H} + \text{CH}_3 \cdot & +72 & 76 & (262) \\
\text{Br} \cdot + \text{H} - \text{CH}_2\text{CH}_3 \rightarrow \text{Br} - \text{H} + \cdot\text{CH}_2\text{CH}_3 & +44 & 55 & (263)
\end{array}$$

homolysis of covalent bonds, or the ionization involved in an S_N1 reaction, the transition state and the products are almost identical, and thus any stabilizing influences (delocalization etc.) which are present in the products will be present to almost the same extent in the transition state, and will therefore have a profound influence on the rate of the reaction.

In reactions where a molecule can be attacked by a reagent in more than one position, or if a reagent reacts with a mixture of two similar compounds, different ratios of products are usually formed under different conditions. As a general rule it is found that the more reactive the reagent, the less selective it is in attacking one molecule (or one position in a molecule) rather than another. Hammond's postulate provides an explanation for these differences in selectivity.

Bromination of toluene by hypobromous acid in aqueous dioxan gives *para* and *meta* bromotoluenes in the ratio 10:1. Bromination may also be carried out more slowly by bromine in acetic acid; here the *para/meta* ratio is increased to 220:1. There is considerable evidence that the brominating agent in the first case is the bromonium ion Br$^+$, or a hydrated form. This is a reactive, high-energy species. Thus reaction (264) will be much more exother-mic than reaction (265) in which molecular bromine is used. Thus the tran-sition state for reaction (264) will resemble the starting materials and the electronic effect of the methyl group will be relatively small. On the other hand

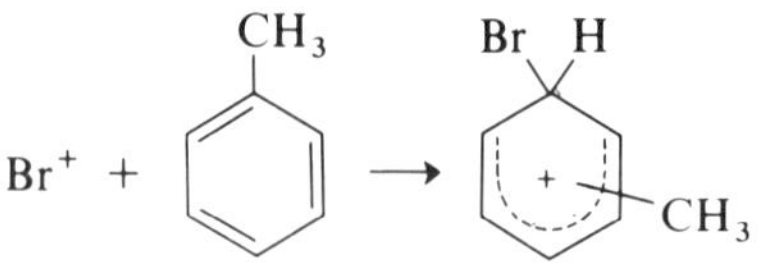

transition state near starting materials.   (264)

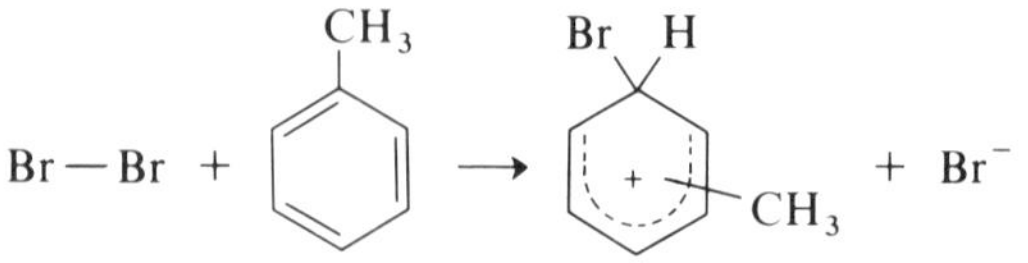

transition state near intermediate **214**.

214

(265)

the molecular bromination (reaction 265) will be endothermic, and the transition state will closely resemble the Wheland intermediate **214**. The methyl group has a much larger stabilizing influence for attack at the *para* than at the *meta* position, and thus the *para/meta* ratio is much higher than for reaction (264).

This state of affairs is quite general: reactive reagents tend to be relatively unselective, whereas less reactive reagents tend to be more selective. If we find that by changing the reaction conditions a substantial change in selectivity occurs, this may mean that a different reacting species is present (e.g. Br^+ rather than Br_2). However more subtle changes may be involved; for example solvents which complex with a reagent often reduce its reactivity and increase its selectivity.

PROBLEMS

6.1. CH_3Br, C_2H_5Br, $i\text{-}C_3H_7Br$ and $t\text{-}C_4H_9Br$ react with iodide ion in acetone with second-order rate constants of relative magnitudes 256 000:1 667:13:1. Suggest an explanation for these differences.

6.2. For the bimolecular reaction of $i\text{-}C_3H_7Br$ with OH^- in aqueous alcohol, the rate constant decreased by a factor of 1·6 when the proportion of water was raised from 20 per cent to 40 per cent. When a corresponding change in solvent was made for the bimolecular reaction of $i\text{-}C_3H_7Br$ with water, the rate increased by a factor of 2·8. Explain these differences.

6.3. Suggest a number of *para* substituents (a) for which σ^+ will be significantly different from σ and (b) for which σ^- will be significantly different from σ.

6.4. 1-Arylethyl acetates decompose on heating in the gas phase in a first-order reaction to give acetic acid and the arylethylene. For substituents on the aromatic ring, a Hammett correlation with σ^+ is observed, with $\rho = -0.7$. What light does this throw on the mechanism of the decomposition?

6.5. It is found that both *cis*- and *trans*-2-arylcylcopentyl toluenesulphonates $ArC_5H_8\cdot O\cdot SO_2\cdot C_6H_4\cdot CH_3$ undergo elimination of toluenesulphonic acid on treatment with potassium t-butoxide to give the arylcyclopentene. The *cis* compounds give a Hammett ρ value of $+1.48$, whereas the *trans* compounds give a ρ value of $+2.76$. Rationalize these results in terms of the likely transition states involved.

6.6. Suggest reactions (other than those discussed in this chapter) which should show (a) positive (b) negative ρ values, and in which correlations would be best (1) with σ, (2) with σ^+, (3) with σ^-; i.e. six categories in all. It may be difficult to find examples of some of these categories.

7. Conclusion

THE previous chapters have illustrated several methods for obtaining information about organic reaction mechanisms. The overall procedure adopted will vary from case to case, but a typical sequence is illustrated in Fig. 7. Normally, the first step is to establish the products of the reaction. A kinetic study may follow, as a result of which a mechanism (or several possible mechanisms) is postulated. Alternatively, ideas about a possible mechanism may emerge from the product study. The postulated mechanism should then be tested, ideally by as many different methods as possible. As a general guide, the more that a proposed mechanism is out of line with mechanisms established for similar reactions, the more rigorously it needs to be tested. Conversely, if a reaction differs only insignificantly from a series of reactions of known mechanism, it is fairly safe to assume that the same mechanism applies with no mechanistic study at all. For example (to reiterate a point made in Chapter 1) if the hydrolysis of n-propyl acetate had been studied rigorously under a particular set of conditions and a mechanism had been established, it would be reasonable to assume that n-butyl acetate hydrolysis under the same conditions would follow the same course, but on the other hand it would be dangerous to assume the same for t-butyl acetate, where there is substitution much nearer the reaction centre. A number of particular tests for mechanism have been described in this book. The list is not meant to be exhaustive, and it should be emphasized that new methods of testing mechanism are continually being invented. The results of the various tests may agree with the proposed mechanism. if they disagree, the mechanism must be refined or a new mechanism chosen. A mechanism which accords with all the known facts may be regarded as an acceptable mechanism.

Further confidence in the mechanism will be gained if the mechanism fits into the general pattern of chemical knowledge, and even more if it links other items of chemical knowledge into a more complete picture. Finally, a mechanism which is 'true' will often suggest other reactions and ideas which can themselves be tested, whereas one which consists of *ad hoc* explanations of individual results will seldom produce ideas of this type.

The Piltdown Man process

It is interesting to look with hindsight on a number of reactions where incorrect ideas about mechanism grew up. We will consider two examples already referred to : the dimerization of triarylmethyl radicals, and the equilibrium between the open-chain and cyclic forms of certain cyclopropane compounds. In each case, features of varying degrees of unlikeliness were observed.

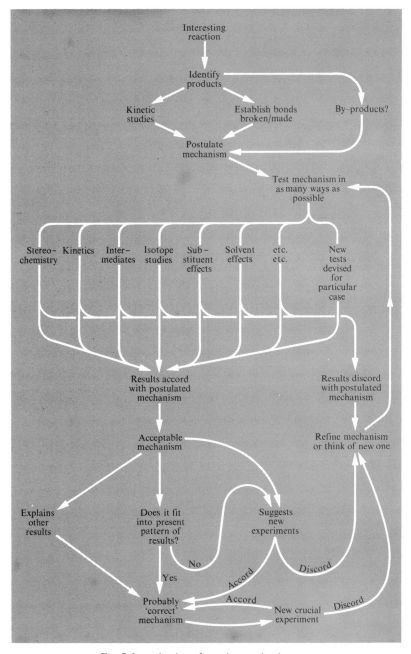

FIG. 7. Investigation of reaction mechanisms.

The triarylmethyl/hexa-arylethane equilibrium never very well explained why there was such a large effect on the equilibrium constant when *three* *para*-substituents were present rather than two. When it was established that the dimerization occurred by α-to-*para* coupling, the reason for the *para* effect was clearly seen to be steric in nature. Formulation of the triphenyl-methyl dimer as hexaphenylethane also led to another difficulty. The rate of formation of triphenylmethylperoxide when the dimer reacts with air is greater than the rate of dissociation of the dimer into free radicals. Accordingly it was assumed that a chain reaction (266) must be involved. However the last stage in this scheme is of a type of which no authentic examples are known :

$$Ph_3C-CPh_3 \underset{\text{fast}}{\overset{\text{slow}}{\rightleftarrows}} 2Ph_3C\cdot \overset{O_2}{\rightarrow} Ph_3COO \overset{Ph_3C-CPh_3}{\longrightarrow} Ph_3C-O-O-CPh_3 + Ph_3C\cdot$$

(266)

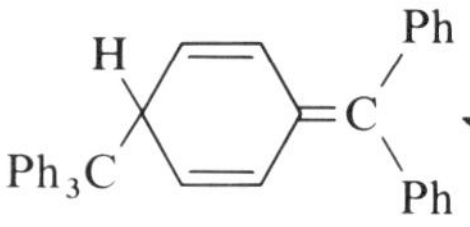

215

(267)

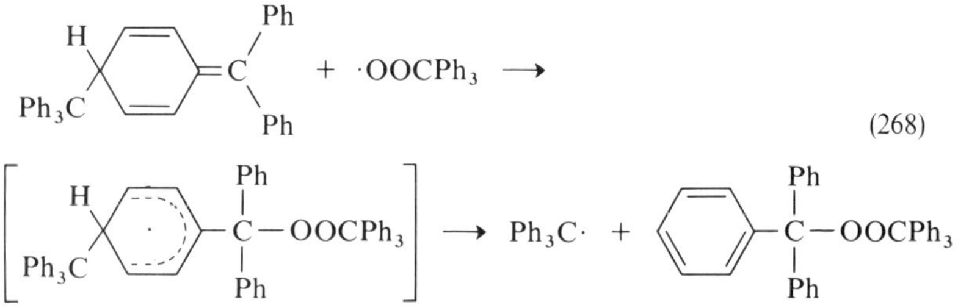

(268)

the direct displacement by one radical of another radical from an acyclic saturated carbon centre. Furthermore in this reaction the carbon atom to be attacked is surrounded by bulky substituents. The reformulation of the dimer as **215** allows the more reasonable mechanistic scheme (267 and 268) to be constructed. The first stage of this sequence is the addition of a radical to a double bond; the second is the reverse of the addition process which takes place in an aromatic substitution reaction. Many reactions analogous to both of these processes are known.

The idea of the Thorpe–Ingold effect arose from the observation that in reactions such as (269) where a cyclopropane ring was thought to be formed, substituents X which tend to produce a large X—C—X bond angle ψ, should contract the C—C—C bond angle θ, and make ring formation easier. Thus the spirocyclohexyl compound **218** (in which the angle was believed at that time to be 120°) apparently underwent ring formation more readily than the dimethyl compound (**216**, X = CH$_3$).

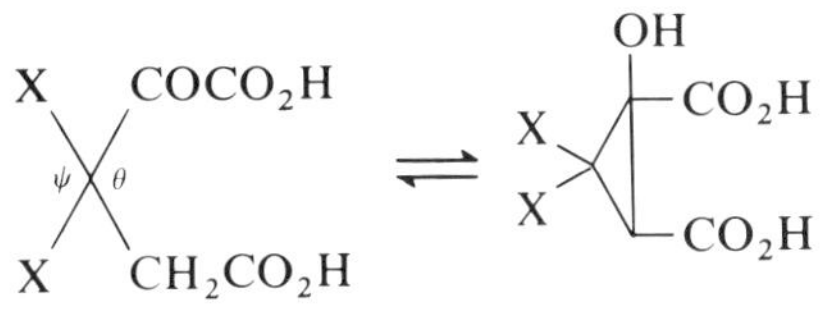

This explanation persisted even after it had been established that the cyclohexane ring is puckered, and that the C—C—C bond angles have the normal tetrahedral value. It was later shown that cyclopropanols are extremely sensitive to heat, which made the formation of such compounds under the conditions of reactions (269) and (270) seem surprising. Thus the stage was well set for the eventual discovery that the compounds originally studied had not been the cyclopropanols at all.

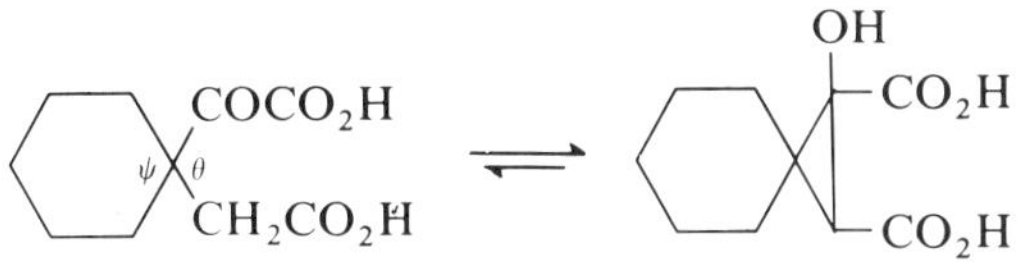

An analogy may be drawn from a different field, palaeontology. In 1912 remnants of an apparently ancient skull were found at Piltdown, Sussex (U.K.). Reconstructions based on this skull indicated that this was a missing-link in the evolution of man. In the light of more recent finds, certain aspects of this skull became more and more puzzling. Finally in 1959 it was proved by radiocarbon dating to be a fake. The acceptance of the skull in the first instance was helped by the fact that this was roughly what had been expected. Later finds made the Piltdown skull appear more and more anomalous, and when it was proved to be a fake this area of human palaeontology was greatly clarified. So it is with reaction mechanisms, which have to be derived within the framework of current knowledge. Outright fakery of results is rare (though several such cases are known to most practising research chemists). More frequently, important results are disregarded as being anomalous. In time the 'anomalies' multiply, and the original theory has to be patched up to account for the new results. Finally, when new results appear, the old mechanism can often be blown away overnight. In spite of all the work which had been done on the 'hexa-arylethane'/triarylmethyl equilibrium, and the fact that most chemists had accepted the ethane structure for the dimer, this whole edifice was swept away by a single short article, with no experimental section.

The theory had become too strained, and everyone was relieved by the simplification afforded by the new structure for the dimer.

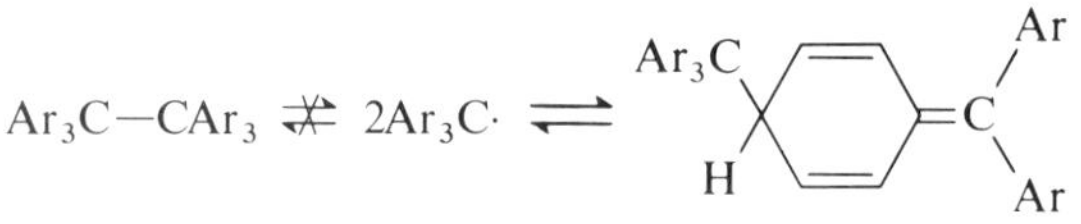

$$Ar_3C-CAr_3 \; \rightleftharpoons\!\!\!\!\!/ \; 2Ar_3C\cdot \; \rightleftharpoons \quad \text{[structure]} \qquad (271)$$

Trendiness

It was suggested in the first section of this chapter that more confidence in a mechanism could be felt if the results in general fitted in with current chemical knowledge. There is an opposite side to the coin. Ideas which do not fit into the established scheme often have difficulty in establishing themselves. Gomberg's original discovery of triphenylmethyl aroused considerable disbelief, since carbon was 'known' always to be tetravalent. When radicals became 'recognized' a celebrated paper by Koelsch, who had prepared the radical **219**, was rejected for publication (though finally published twenty-five years later) on the grounds that radicals were 'known' to react with oxygen, although **219** was stable in the air.

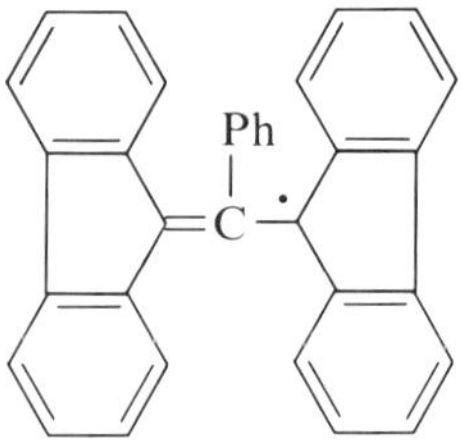

219

It is probably right that the really new mechanistic ideas should require more detailed and adequate support than ideas which are only trivial extensions of existing ones. At any rate this is usually what happens.

Conclusion

In this book, various techniques which throw light on organic reaction mechanisms have been described. A choice of material has to be made: many interesting techniques and concepts have been left out. The restriction of the field to homogeneous organic reactions has also forced the omission of many other important and interesting reactions. However, it is hoped that the approach used in this book will be found useful by those who wish to go on to more advanced study of mechanism in particular areas, not only of 'pure organic' reactions, but also of polymer reactions, reactions at surfaces, inorganic reactions, and biochemical reactions. Each of these areas has evolved techniques which are additional to, or replace some of, those described in this book.

In tackling new areas, William of Occam's sixteenth-century advice is still valid: do not multiply entities (concepts) unnecessarily. Various new concepts arise at intervals, for example non-classical carbonium ions and the various kinds of ion-pair interactions in organic chemistry, gel effects in polymerizations, and the ideas of enzyme/substrate interactions in biochemistry. Before you accept any new concept, you should ask: What is the evidence in favour of it? Could the facts be explained in a simpler way without use of the concept? Only if satisfactory answers to both these questions can be given, should the new concept be accepted.

PROBLEMS

7.1. Phenyl diazonium chloride reacts with the azide ion to give phenyl azide and nitrogen.

$$\underset{1 \quad 2}{\overset{+}{Ph-N}\equiv N} + \underset{3 \quad 4 \quad 3}{\overset{-}{N}=\overset{+}{N}=\overset{-}{N}} \rightarrow \underset{a \quad b \quad c}{\overset{+}{Ph-N}=N=\overset{-}{N}} + \underset{d \quad d}{N\equiv N}$$

(a) Suggest a mechanism for this reaction.
(b) It is found that if the phenyl diazonium chloride is labelled with ^{15}N at atom (1), all the ^{15}N turns up at (a) in the phenyl azide. Revise or change your first postulated mechanism (if necessary) to take account of this.
(c) If the phenyldiazonium chloride is labelled at position (2), 83 per cent of the ^{15}N is found at (b) in the phenyl azide and the remainder is present in the nitrogen evolved. No ^{15}N is found at (a) or (c). Modify your mechanistic postulate again to account for these additional data (it may be necessary to consider two competing reactions). What further experiments can you suggest to test your mechanism?

7.2. For students who have an interest in reaction mechanisms in another field, e.g. inorganic chemistry, biochemistry, polymer chemistry, or surface chemistry. Are all the techniques described in this book relevant to the other field? What other techniques are available or particularly suitable for the study of reaction mechanisms in your chosen area?

7.3. Collect together (from this book or elsewhere) as much evidence as possible for the mechanism of a particular reaction (e.g. the Claisen rearrangement). Does the evidence, considered as a whole, seem convincing? What other experiments, if any, suggest themselves?

Hints for solving problems

1.1. Numbers refer to points made on pages 4 to 10.
(a) 1,3 (b) 3,4 (c) 1 (d) all right (e) 5 (f) 6 (g) 5.

2.1. Isotopes.

2.3. How would you establish the presence of chlorine in any reaction mixture?

2.6. The phenyl compound reacts very much faster than the methyl compound.

3.2. Consider uni- and bi-molecular reactions of the mercurials as the rate-determining processes.

3.3. Two of the reactants may form an intermediate which then rapidly attacks the remaining component.

3.4. See p. 40 for treatment of a somewhat analogous reaction.

3.6. See Table 3, p. 46.

4.1. A rearrangement in the carbon skeleton has taken place in one of the products.

4.2. Think of ways in which 1,2,3-triphenylpropane could have been formed from bibenzyl.

4.3. Sodium is much more electropositive than magnesium.

4.4. MnO_2 must have been formed by an oxidizing agent not present in the original solution (i.e. by an intermediate). Suggest one or more plausible intermediates which might oxidize Mn(II).

4.5. A common intermediate normally gives similar product ratios.

5.1. One reaction must involve retention, the other inversion.

5.2. Deuterium.

5.3. Consider the hydroboration reaction.

5.4. The normal 'ionic' elimination pathway occurs with *trans* stereochemistry.

5.5. Radical dissociation, or a molecular reaction involving the NC group 'turning round' without becoming free?

6.1. Consider steric influences on the transition state for an S_N2 reaction.

6.2. See Table 8, p. 114.

6.3. Consider substituents which will be able to delocalize positive and negative charge.

6.4. The ρ value and the correlation with σ^+ should suggest how much of what kind of charge is accumulating at the 1-position during the reaction.

6.5. More negative charge develops at the 2-position in the *trans* compound.

7.1. Some of the possible reactions are:
(a) direct displacement of N_2 from the phenyl diazonium chloride by N_3^-

(b) attack by N_3^- at nitrogen atom (2) to give $Ph-N{=}N-N{=}\overset{+}{N}{=}\overset{-}{N}$ **A** as an intermediate

(c) the formation of **A** which then cyclizes to give $\text{Ph}-\text{N}\begin{array}{c}\diagup\text{N}=\text{N}\\ \\ \diagdown\text{N}=\text{N}\end{array}$ **B**. **A** and **B**

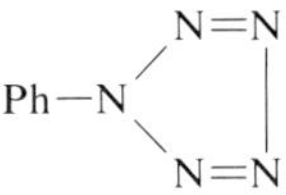

could then decompose to give the products.

Route (a) is excluded by one piece of evidence in the question (which?). Try to account for the whole of the evidence in terms of (b) and (c).

In suggesting other possible experiments, remember that **A** and **B** are postulated *intermediates*.

7.3. The index and bibliography should help here.

Answers to problems

1.1. (a) As written, this is a quadrimolecular reaction. This mechanism also does not explain the role of the $Ni(CN)_2$.

(b) The reverse of this reaction would be a quinquemolecular reaction. Hence by the principle of microscopic reversibility, the forward reaction will also involve a different course.

(c) This reaction scheme seems rather complex: additionally it does not explain why products are formed almost exclusively from the last radical in the sequence, rather than say the one before, or the one further down the chain. The alternative mechanism.

is simpler, and provides an explanation for the specificity.

(d) Nothing wrong here.

(e) A molecular addition of this type with both bonds forming at once should form the two bonds on the same face of the acetylene molecule. This would give the opposite geometrical isomer.

(f) The ionization of benzene in solution of acetic acid is extremely unfavourable energetically.

(g) The intermediate nitrogen radical postulated has nine electrons in its valence shell.

2.1. $CD_3CH_2Cl + OH^- \longrightarrow CD_3CH_2OH + Cl^-$
$^{14}CH_3CH_2Cl + OH^- \longrightarrow {}^{14}CH_3CH_2OH + Cl^-$

2.2. (a) Not impossible, but the fact that approximately 50 per cent of each of the two products is formed under a variety of reaction conditions is strong evidence that the result is not just due to a mere coincidence.

(b) If a carbonium ion is formed, and a hydride shift follows as shown, one would expect at least some product(s) from further hydride shifts which should be equally easy.

The formation of carbonium ions in a basic liquid such as liquid ammonia is almost inconceivable.

2.3. Two ways which have been used are described in Chapter 4, p. 77. Many alternatives are feasible.

2.4. E.g. Isolate the methanol, look for the C–D stretching band in the i.r. Alternatively, react the methanol chemically to give a derivative from which the hydrogen (or deuterium) attached to oxygen has been lost, for example methyl acetate. The i.r. or mass spectrum could then be compared with an authentic non-deuteriated sample.

2.5. The fact that **A** is oxidized whereas **C** is not suggests that the oxidation involves the nitrogen-bound hydrogen in **A**. Hence **B2** is the more likely structure for the product. Supporting evidence: **B** is yellow coloured, an indication that the conjugated —N=N—Ph group in **B2** is present. **B1** contains no group which is likely to impart a yellow colour to the molecule. **B1** would have an N—H group and **B2** an O—H group. These should be distinguishable on the basis of their i.r. or n.m.r. spectra.

2.6 No. Virtually all of the phenyl compound would have reacted in a period of time when very little of the methyl compound had decomposed. Hence there would be almost no chance for crossover reactions to occur.

3.1. Ideas include (a) Titration of OH^- left or of Br^- formed. (b) Pressure change (or volume change at constant pressure). (c) Spectrophotometric determination of remaining bromine.

3.2. (a) $(PhCH_2)_2Hg \rightarrow PhCH_2CH_2Ph + Hg$

   or $\rightarrow PhCH_2\cdot + \cdot HgCH_2Ph \rightarrow Hg + \cdot CH_2Ph$

   or $\rightarrow 2PhCH_2\cdot + Hg$

   followed by $2PhCH_2\cdot \rightarrow PhCH_2CH_2Ph$

   (b) $\begin{matrix} Me_3Si-Hg-SiMe_3 \\ Me_3Si-Hg-SiMe_3 \end{matrix} \rightarrow \begin{matrix} Me_3Si & Hg-SiMe_3 \\ | & + & | \\ Me_3Si & Hg-SiMe_3 \end{matrix} \rightarrow \begin{matrix} Hg & SiMe_3 \\ + \\ Hg & SiMe_3 \end{matrix}$

   or $\rightarrow \begin{matrix} Me_3Si & \cdot HgSiMe_3 \\ | & + \\ Me_3Si & \cdot HgSiMe_3 \end{matrix} \rightarrow 2Hg + 2Me_3Si\cdot$

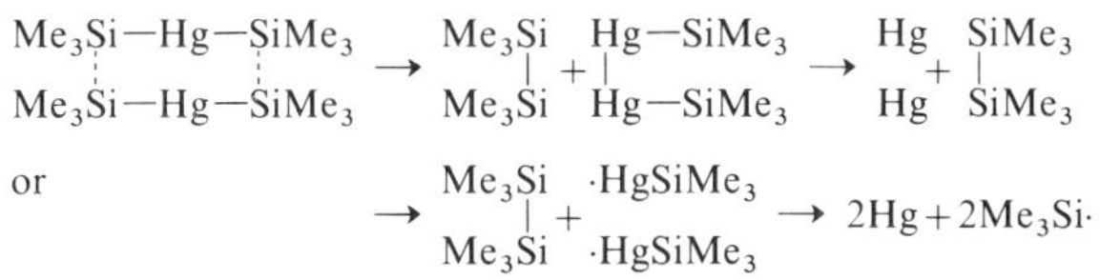

   followed by $Me_3Si\cdot + Hg(SiMe_3)_2 \rightarrow Me_3Si-SiMe_3 + Hg + \cdot SiMe_3$

   $2Me_3Si\cdot \rightarrow Me_3Si-SiMe_3$

   Steady-state treatment may be used to show that this second scheme also gives second-order kinetics.

3.3. An intermediate must be formed in a slow step involving nitric acid and possibly acetic acid. This reactive intermediate (believed to be the nitronium ion) is removed by a fast reaction with the benzene. Hence the rate of nitration is the rate of formation of the intermediate nitronium ion and the benzene concentration is unimportant.

3.4. Rate $= (k_1/k_4)^{\frac{1}{2}}k_3[\text{Initiator}]^{\frac{1}{2}}[\text{RH}]$
   $RO_2\cdot$ is present in much higher concentration than $R\cdot$.

3.5.

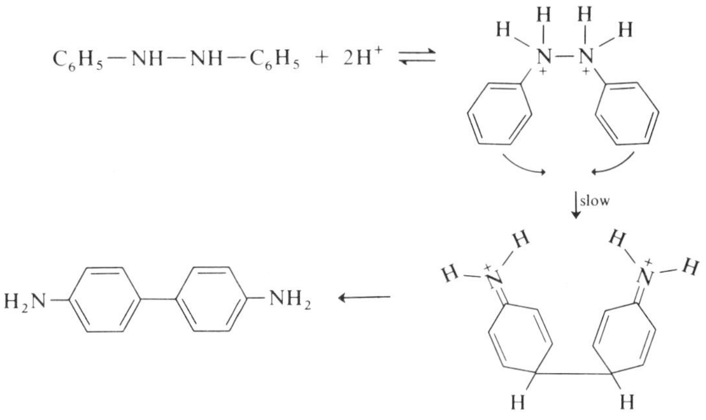

The absence of a deuterium isotope effect precludes any mechanism in which breakage of the p-C–H bonds is kinetically significant. This mechanism accounts fairly well for the facts listed in the question, but at the time of writing the detailed mechanism of this reaction is still uncertain.

3.6. The molecular reaction (a). The radical decomposition (b) should have a considerably higher A-factor.

4.1. $PhCHMe-CHMeNH_2 \xrightarrow{HNO_2} PhCHMe-CHMe^+$. Migration of a methyl group or hydride, followed by attack by acetic acid gives two of the products, phenyl migration or no migration gives the third. If the radical or carbanion had been involved, these rearrangements would not have taken place.

4.2. $(PhCH_2)_2Hg \xrightarrow[\text{stages}]{1 \text{ or } 2} Hg + 2PhCH_2 \cdot \rightarrow PhCH_2CH_2Ph$
$PhCH_2CH_2Ph + \cdot CH_2Ph \rightarrow PhCH_3 + Ph\dot{C}H-CH_2Ph \xrightarrow{PhCH_2 \cdot}$
$PhCH_2-CHPh-CH_2Ph$.
Add bibenzyl to the reaction mixture: this should increase the amount of triphenylpropane formed. Addition of a compound, such as nitrobenzene, which reacts with benzyl radicals, should reduce the amount of bibenzyl formed.

4.3. $CH_3 \frown H \frown CH_2 \frown CH_2 \frown OC_2H_5 \rightarrow CH_4 + CH_2=CH_2 + {}^-OC_2H_5$
Since Grignard reagents do not do this reaction, they must be less ionic than organosodium compounds. Hence Grignard reagents must have considerable covalent character in the C–Mg bond.

4.4. An intermediate must be formed which is a stronger oxidant than Cr(VI) or an aldehyde. Cr(IV) has been suggested.

4.5. Since the product ratios in the two reactions are different, a common intermediate is unlikely. As $PhCHN_2$ gives PhCH :, the other reactant system probably does not, and direct reaction of PhCHBrLi with 1-butene is likely.

5.1. The second reaction does not involve the Si atom, and hence occurs with retention. Thus the first reaction involves inversion. See also p. 85.

5.2. Use the optically active deuterium compound CH_3CHDBr.

5.3. $C_3H_7-CH{=\!=}CH-CH_3 \rightarrow C_3H_7-CH=CH-CH_3 \rightarrow$
$(C_6H_{13})_2B{-}{-}{-}{-}H \quad +(C_6H_{13})_2B-H$

$$C_3H_7-CH_2-CH-CH_3 \quad \text{etc.}$$
$$\underset{B(C_6H_{13})_2}{|}$$

A double bond cannot be formed at a quaternary carbon atom. Hence the boron cannot pass such a grouping during this rearrangement.

5.4. Elimination of hydrogen halides by base occurs most readily when the hydrogen and the halogen are *trans* to each other. Elimination to give **183** would have to be *cis* elimination.

5.5. Thermal dissociation to Ph Me Et C· and CN· occurs, followed by combination of the radicals to give Ph Me Et C·CN. Since a (planar) radical intermediate is formed, racemization is expected. If the N≡C group merely 'turned round' in a molecular reaction, the configuration would be retained. Optically active Ph Me Et C·CN is not racemized under the conditions of this experiment, thus ruling out isomerization followed by racemization.

6.1. In the transition state for an S_N2 reaction, five groups are coordinated to the central carbon atom as opposed to four in the original halide. Hence steric effects are more important in the transition state, and the more alkyl groups that are attached to the central atom, the greater this effect will be.

6.2. The first reaction involves a spreading of charge in the transition state, the second involves creation of charge. Thus the first will be decreased in rate on increasing the ionizing power of the solvent, the second will be increased (by a larger amount).

6.3. σ_p^+ : NMe$_2$ NH$_2$ OMe   SMe   OH   Cl
     σ_p^- : CHO   COMe   CO$_2$Me   CN   NO$_2$

6.4. Positive charge is developed at the 1-position but the small value of ρ suggests that the molecule has not ionized. A molecular transition state with a certain amount of charge development seems likely.

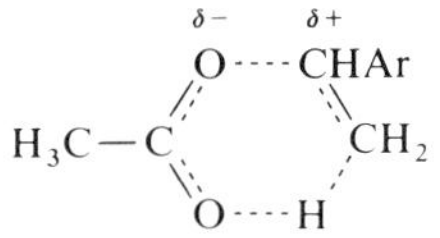

6.5. Concerted elimination of toluenesulphonic acid can take place most easily if the H and O·SO$_2$·C$_6$H$_4$·CH$_3$ groups are *trans* to each other (i.e. in the *cis* compound). In a concerted process relatively little carbanion character is developed at the 2-position. In the *trans* compound where the groups are not so favourably disposed for concerted elimination, there is a greater development of carbanion character in the transition state.

7.1. Direct displacement of N$_2$ by N$_3^-$ is precluded by the labelling experiment at nitrogen atom (1). The following scheme accounts for the distribution of ^{15}N from the (2) position.

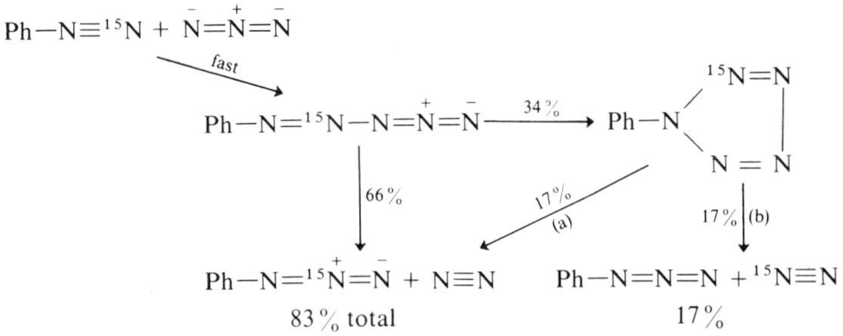

If the reaction is carried out at $-50°C$, about 75 per cent of the N$_2$ is evolved in a first order reaction (no enrichment of ^{15}N in the N$_2$ liberated). On warming to $0°C$, the remainder of the N$_2$ is evolved, again in a first order reaction, with ^{15}N enrichment in accord with the scheme.

Bibliography

All the general books (below) except that of Weissberger treat reaction mechanism from the viewpoint of the type of reaction. They are listed in an approximate order of increasing 'difficulty'. Most give references to reviews and original scientific papers. The references given to particular topics are not exhaustive and do not necessarily identify the key workers in an area. However, these papers should allow earlier papers on the same subject to be located.

General Books

P. SYKES (1970) *A guidebook to mechanism in organic chemistry*. Longmans, London.
E. S. GOULD (1959) *Mechanism and structure in organic chemistry*. Holt, New York.
J. HINE (1962) *Physical organic chemistry*. McGraw-Hill, New York.
C. K. INGOLD (1969) *Structure and mechanism in organic chemistry*. Bell, London.
J. MARCH (1968) *Advanced organic chemistry: reaction, mechanisms, and structure*. McGraw-Hill, New York.
R. W. ALDER, R. BAKER and J. M. BROWN (1971). *Mechanism in organic chemistry*. Wiley, London.
E. M. KOSOWER (1968) *Introduction to physical organic chemistry*. Wiley, New York.
A. WEISSBERGER (ed.) (1961) *Techniques of organic chemistry. Vol. VIII. Investigation of rates and mechanisms of reactions*. Interscience, New York.

Chapter 1

Optimization. J. D. ROSE (1971) *Chem. Br.* **7**, 421.
Rocket fuels. R. M. LAWRENCE and W. H. BOWMAN (1971) *J. chem. Educ.* **48**, 458.
The reaction of benzaldehyde with acetophenone. E. COOMBS and D. P. EVANS (1940) *J. chem. Soc.* 1295.
Benzyne and radical reactions—see Chapter 4.

Chapter 2

The triphenylmethyl dimer problem. H. LANKAMP, W. TH. NAUTA and C. MacLEAN (1968) *Tetrahedron Lett.* 249; W. THEILACKER and M.-L. WESSEL-EWALD (1955) *Annalen.* **594**, 214.
The Thorpe-Ingold Effect. K. B. WIBERG and H. W. HOLMQUIST (1959) *J. org. Chem.* **24**, 578.
The Beckmann rearrangement. N. V. SIDGWICK (1969) *The Organic Chemistry of Nitrogen*. Clarendon Press, Oxford.
Ester hydrolysis, see Chapter 4, tetrahedral intermediates.
Reactions involving carbonium ions, carbanions, radicals, benzynes, see Chapter 4.

Chapter 3

A. A. FROST and R. G. PEARSON (1961) *Kinetics and Mechanism*. Wiley, New York.
G. L. PRATT (1969) *Gas kinetics*. Wiley, London.
E. F. CALDIN (1964) *Fast reactions in solution*. Blackwell, Oxford.
R. HUISGEN (1970) Kinetic evidence for reaction intermediates: in *Angew. Chem.* (International Edn. in English) **9**, 751.
S_N1 and S_N2 Reactions. INGOLD's 'Structure and Mechanism'—see general section.
The hydrogen iodine reaction. J. H. Sullivan (1967). *J. chem. Phys.* **46**, 73; R. HOFFMANN (1968) *J. chem. Phys.* **49**, 3739.

The bromine chloroform reaction. J. H. SULLIVAN and N. DAVIDSON (1951) *J. chem. Phys.* **19**, 143.

Cyclobutane decomposition, C. T. GENAUX, F. KERN, and W. D. WALTERS (1953) *J. Amer. chem. Soc.* **75**, 6196; S. W. BENSON (1961) *J. chem. Phys.* **34**, 521.

Isotope effects. R. P. BELL (1959) *The proton in chemistry*. Methuen, London. In aromatic substitution; L. MELANDER (1949) *Nature, Lond.* **163**, 599.

Chromate oxidations. F. H. WESTHEIMER (1949) *Chem. Rev.* **45**, 419; F. HOLLOWAY, M. COHEN, and F. H. WESTHEIMER (1951) *J. Amer. chem. Soc.* **73**, 65.

Chapter 4

Styrene/methyl methacrylate polymerization. C. WALLING, E. R. BRIGGS, W. CUMMINGS, and F. R. MAYO (1950) *J. Amer. chem. Soc.* **72**, 48.

Carbonium ions. D. BETHELL and V. GOLD (1967) *Carbonium ions*. Academic Press, London; P. DE MAYO (ed.) (1963) *Molecular rearrangements, Vol. I.* Interscience, New York; R. H. DEWOLFE and W. G. YOUNG (1956) *Chem. Rev.* **56**, 753.

Radicals. W. A. PYROR (1968) *Free radicals*. McGraw-Hill, New York; W. A. PYOR (1966) *Introduction to free radical chemistry*. Prentice-Hall, New Jersey. Fluorination of methane. E. H. HADLEY and L. A. BIGELOW (1940) *J. Amer. chem. Soc.* **62**, 3302. Cumene autoxidation. J. R. THOMAS (1963) *J. Amer. chem. Soc.* **85**, 591. Autoxidation of organoboron compounds. A. G. DAVIES, K. U. INGOLD, B. P. ROBERTS, and R. TUDOR (1971) *J. chem. Soc.* (*B*) 698. Use of CIDNP, R. W. JEMISON and D. G. MORRIS (1969) *Chem. Commun.* 1226.

Carbanions. D. J. CRAM (1965) *Fundamentals of Carbanion Chemistry*. Academic Press, New York. Decarboxylation of pyridine-1-carboxylic acid. M. R. F. ASHWORTH, R. P. DAFFERN, and D. Ll. HAMMICK (1939) *J. chem. Soc.* 809. Re-arrangement of lithium salts of esters. U. SCHÖLLKOPF and D. WALTER (1961) *Angew. Chem.* **73**, 545.

Carbenes and Benzynes. T. L. GILCHRIST and C. W. REES (1969) *Carbenes, Nitrenes, and Arynes*. Nelson, London.

¹⁴C-Benzyne. J. D. ROBERTS, D. A. SEMENOW, H. E. SIMMONS, JR., and L. A. CARLSMITH (1956) *J. Amer. chem. Soc.* **78**, 601.

Competition between furan and cyclohexadiene for benzyne. R. HUISGEN and R. KNORR (1963) *Tetrahedron Letters*, 1017.

Tetrahedral intermediates. ¹⁸O in ester hydrolysis. D. N. KURSANOV and R. V. KUDRYAVTSEV (1956) *J. gen. Chem. U.S.S.R.* **26**, 1183; M. L. BENDER (1960) *Chem. Rev.* **60**, 53.

Ring closure of a tetrahedral intermediate. H. E. ZAUGG, V. PAPENDICK, and R. J. MICHAELS (1964) *J. Amer. chem. Soc.* **86**, 1399.

Chapter 5

E. L. ELIEL (1962) *Stereochemistry of carbon compounds*. McGraw-Hill, New York.

G. HALLAS (1965) *Organic stereochemistry*. McGraw-Hill, London.

Silicon stereochemistry. L. H. SOMMER and co-workers (1961) *J. Amer. chem. Soc.* **83**, 2210.

Cis elimination. D. H. R. BARTON, A. J. HEAD, and R. J. WILLIAMS (1952) *J. chem. Soc.* 453.

Hydroboration. H. C. BROWN (1962) *Hydroboration*. Benjamin, New York.

Bi-radical intermediates. L. K. MONTGOMERY, K. SCHUELLER, and P. D. BARTLETT (1964) *J. Amer. chem. Soc.* **86**, 622.

Orbital symmetry. R. B. WOODWARD and R. HOFFMANN (1970). *The conservation of orbital symmetry*. Verlag, Weinheim. H. E. ZIMMERMAN (1971) *Acct. Chem. Research* **4**, 272.

Chapter 6

Hammett Equation. P. R. WELLS (1963) *Chem. Rev.* **63**, 171. σ^+. H. C. BROWN and Y. OKAMOTO (1958) *J. Amer. chem. Soc.* **80**, 4979.

Ionization potentials of benzyl radicals. A. G. HARRISON, P. KEBARLE, and F. P. LOSSING (1961) *J. Amer. chem. Soc.* **83**, 777.

N-nitroamines. W. N. WHITE and J. R. KLINK (1970) *J. org. Chem.* **35**, 965.

N-bromosuccinimide. R. E. PEARSON and J. C. MARTIN (1963) *J. Amer. chem. Soc.* **85**, 354.

Electrophilic aromatic substitution. R. O. C. NORMAN, and R. TAYLOR (1964). *Electrophilic substitution in benzenoid compounds*. Elsevier, Amsterdam.

Pyrolysis of alkyl halides. A. MACCOLL (1969) *Chem. Rev.* **69**, 33.

Pyrolysis of arylethyl acetates. R. TAYLOR and G. G. SMITH (1963). *Tetrahedron* **19**, 937.

Solvent effects. See in particular the books by INGOLD and KOSOWER in the general section.

Chapter 7

The azide/diazonium chloride reaction (problem). I. UGI, R. HUISGEN, K. CLUSIUS, and M. VECCHI (1956) *Angew. Chem.* **68**, 705, 753.

Index

Reactions have as far as possible been classified under **addition**, **elimination**, **rearrangement**, and **substitution**. General terms (e.g. halogenation) have been used rather than specific terms (e.g. bromination). Few entries have been made under the names of compounds or classes of compounds. A page entry in italics indicates a definition or explanation of the term. References to problems are indicated by a letter P after the page number.